RESEARCH & THE ANALYSIS OF RESEARCH HYPOTHESES

Volume 2

KATHLEEN THOMAS ALLAN, PH.D.

Print information available on the last page.

Rev. date: 04/18/2016

To order additional copies of this book, contact:
Xlibris
1-888-795-4274
www.Xlibris.com
Orders@Xlibris.com

DEDICATED TO

Anne Russo, friend and mentor. Thank you, for your continued encouragement. It has made all the difference.

Thanks also to Dorothy Shields & Amita for your help in getting started.

Credits and acknowledgment of other sources with permission information is included as footnotes in the text at the location of the source.

Internet URLs appear in the text at appropriate points. Every effort has been made to ensure accuracy; however, internet sites frequently change which may include a change to the URL. Using a search engine will usually provide alternate sources of the required information.

PREFACE

The design of this volume follows the same format as Volume 1, owing a great deal to concepts for designing curriculum researched by M. David Merrill, Ph.D., as a professor at Brigham Young University, in the 1970s. It uses the principles of Rule – Example - Practice. The term "rule" covers such items as a mathematical rule, a classification paradigm, a descriptive category, or other information that gives the students a "rule" for understanding the main concept being taught. The term "example" covers the criteria showing how the rule worked. The term "practice" gives the student an example of how to use the rule.

The two volumes of *RESEARCH & THE ANALYSIS OF RESEARCH HYPOTHESES* provide a basis for doing a research study which graduate students can use as a model for their thesis or dissertation studies. Volume 1 covers basic principles and processes for doing a research study and Volume 2 covers the five major procedures for testing research data: the z-test, the t-test, the Pearson Product Moment Correlation test, the Spearman Rank Correlation test and the Chi Square test. Note:

'Chi' is pronounced "Kai" (rhymes with "sigh").

The 'Chi-Square' symbol is: $\pmb{\chi^2}$

Merrill's concepts have had particular impact on *Volume 2*. This volume shows practical methods for analyzing research data. Merrill's "rule" is described and defined in the "**Purpose:**" portion of each unit of instruction. The "example" and matching "practice" are described in the portion labeled "**Objectives:**" The body of the unit describes the rule in more detail giving examples as needed. Finally, there is an "assignment" which requires the student to put the rule into practice.

OUTLINE - VOLUME 2:

TERMINAL OBJECTIVE FOR THE COURSE:

Write a report in the form of an article suitable for publication, reporting your research and the results, with a minimum standard of a B grade. Your Instructor may determine a different grade requirement.

Volume 2 gives you the analysis tools for determining the results and the significance of your research and how to finalize your report.

SPECIFICATION FOR REPORTS, GROUP DISCUSSIONS, & ASSIGNMENTS:

All Reports should be typed and double spaced. The reports for this course should be hand-in-hand with your research notebook and the group discussion activities. ***Make sure spelling and grammar are correct.***

- One major purpose of the group discussion is to enable each person in the class to improve their work.
    - When critiquing, be honest but not unkind.
    - No put-downs or derogatory remarks.
    - Be helpful with suggestions for improvement.

ONLINE ACTIVITIES:

As in Volume 1, URLs are given about such items as thesis and dissertation writing and analysis of data. Take advantage of these as the various sites (and others) will be helpful when you begin to do other studies. {Note: These are accurate at the time this volume goes to publication. Some of these are stable, but others may change as the owners of the site make internal changes. This is inevitable. If this happens in this course, your instructor should be able to assist you to find others. Share information with other members of the course if you find any URLs without assistance.

RESEARCH NOTEBOOK OR DIARY:

Continue to **keep notes of what happens as you progress through this course.** Any insights about your research, or unexpected problems, or if you discover something you hadn't thought of will be really useful when you start working on your final report.

ASSIGNMENT WORKBOOK/COMPUTER FOLDER:

Keep your workbook/computer folder up to date. You will often be required to prepare handouts for the rest of the students in your class so they can review your work as part of a class discussion. Arrange for copying with your instructor.

Contents

SECTION 4: PROCEDURES FOR ANALYZIING DATA

PURPOSE OF SECTION 4:

A statement made in Volume 1, Section 1: *Formal research is the process of collecting, organizing, analyzing, and interpreting data. This process is generally referred to as **hypothesis testing**. The purpose of testing one's hypotheses is to gain insight increase understanding of the research question, and present the results to our colleagues. This should be regarded as the essential part of doing research. It is no use deciding on your hypotheses or gathering data if you don't analyze for significance. You need to be able to report whether your hypotheses did or did not provide significant insight into your research question.*

While it is most often the case that significance is found, there are times when data does not show significance, but never-the-less it is possible to gain insight even when there is disappointment. Such questions as: Why didn't it work? What was missing from the theory? Was the hypothesis focused too tightly or not tightly enough on the real question? How could the question be changed to more effectively guide the research? Finding that the data supports the null hypothesis is NOT a signal that the research should be tossed in a circular file but that the researcher needs to determine why the data showed there was no significance in the research question as it is currently defined.

OBJECTIVES OF SECTION 4:

- Describe each of the following tests with regard to the type of data that can be analyzed, when you would use that test, and what type of results you would expect.
  - The z-test.
  - The t-test.
  - Pearson Correlation test.
  - Spearman Rank Order Correlation test.
  - The Chi Square test.

- Use the above five tests to determine if you can reject or must fail to reject the null hypothesis.

- Explain how the concepts of "standard error" and "degrees of freedom" are of value in performing certain analytical tests.

- Use the four step plan to prepare your final report:

- Examine your data and with the help of Appendix A determine what graphics will be of value to illustrate the data.

- Complete your final report.

A BRIEF REVIEW OF THE RESEARCH PROCESS (Vol 1):

This flow chart reviews the main concepts discussed in Volume 1. Highlights of the concepts are:

- The literature search for your field of study, enabling you to determine what has already been done in the area of your problem statement.

- Your hypotheses to present your theory about your questions, which guides your study.

- You also made a research proposal and collected appropriate data.

The next step is to analyze your data.

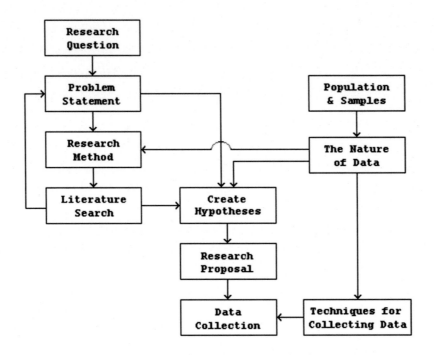

Figure 1: Concepts covered in Volume 1.

THE ANALYSIS PROCESS (VOL 2):

Testing your data enables you to accept or reject your hypotheses. You must examine the results of your study and form conclusions which should be a reasoned analysis of your research. The three types of data are analyzed with different tests.

It is helpful to create some graphics which will represent the data and the results of your study. The results are combined and assist you in stating your conclusions. The final report presents the results of your study to your colleagues and hopefully will expand the knowledge base in your field.

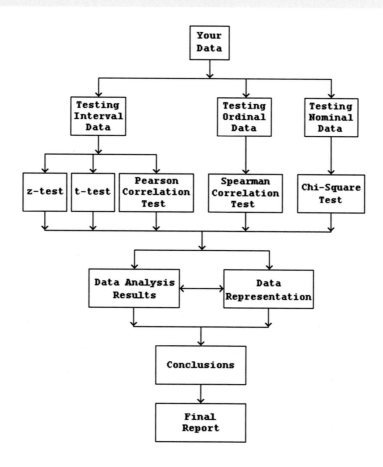

Figure 2: Concepts for analyzing data (Volume 2).

ASSISTANCE IN YOUR THESIS OR DISSERTATION STUDIES:

When doing a research study, you are not alone. The university statistics department like their graduate students to gain experience in using statistical analysis, so it will be possible for you, when analyzing your study data, to consult with one of their graduate students **but you should make arrangements in advance**. In consulting with the statistics department, it will be important to understand the basics of research analysis so you can make sure you can talk intelligently to the graduate student. Even if you cannot remember the analytical tests themselves, you need to understand the purpose of each of the tests you might have to use and to guide your graduate student helper in how the analysis should be conducted.

UNIT 4A: TESTING HYPOTHESES

PURPOSE:

Now we come to the essential part of doing research: analyzing for significance. Your hypothesis testing must be done in a orderly manner. Harvey Berman[1] has come up with a very useful plan for analyzing data in a four step process. If you follow these steps, you will optimize your research.

This section also reviews basic principles from Volume 1, essential for the analysis of hypotheses, with an expansion of the information given to show how the principles fit into the analysis process. These principles include: parameters vs. statistics; how the nature of data impacts the analysis process; how the format of your hypotheses determines how the hypothesis is tested; types of errors and various other factors that may affect your analysis; the concept of tails on a curve; one-tail and two-tail curves; regions of acceptance and rejection of the null hypothesis; and algorithms for calculating mean, variance and standard deviation.

OBJECTIVES:

- Describe each of the four steps for testing hypotheses in Harvey Berman's plan and how these four steps will be of value in **your** research project.

- Formally assess basics of the analysis process to determine how they fit into the recommended plan.

- Describe the decision rules that enable you to reject or fail to reject the null hypothesis.

- Explain the major types of error that a researcher may make when analyzing data.

- Explain how the concepts of "standard error" and "degrees of freedom" are of value in performing certain statistical tests.

- Explain how you can know whether or not you can reject the null hypothesis.

- Use the discussed tests to analyze your own data.

- Complete the four steps for testing your hypothesis.

---

1 This plan is based on information in the AP Tutorial Lesson 5: and is used with the permission of the author Harvey Berman. Refer to:
http://stattrek.com/Lesson5/HypothesisTesting.aspx?Tutorial=APhttp://stattrek.com/Lesson5/HypothesisTesting.aspx?Tutorial=AP

A USEFUL PLAN FOR TESTING HYPOTHESES:

A broad generalization of the null hypothesis is that the data shows that the actions taken in the study made no difference or that the results of the various parts of the study are the same. Harvey Berman has written a series of online lessons for those who plan to obtain AP credit for college subjects. One very useful plan he has set up for the analyses of data is a four step process for testing hypotheses. If you follow these steps, you will be optimizing your research:

1. ### STATE THE HYPOTHESES:

   State the null and alternative hypotheses which must be mutually exclusive, that is, if one is true, the other must be false. This step is essential since it is the null hypothesis (H_0) that you need to test but having the alternative hypothesis (H_a) clarifies what you are hoping to achieve, in other words, that actions taken in the study **do** make a difference. NOTE: H_0 always includes an equality: less than or equal to ($\leq$), greater than or equal to ($\geq$), or equal to ($=$). H_a always excludes the equality (less than $<$, greater than $>$, or not equal to $\neq$).

2. ### FORMULATE AN ANALYSIS PLAN:

   As part of the analysis plan, describe which tests you will apply to your data in order to accept or reject the null hypothesis. The evaluation plan usually focuses on a single *test statistic* (mean score, proportion, t-score, z-score, etc.) In discussing each test in this unit, criteria will be given to help you to know which test to choose.

3. ### ANALYZE SAMPLE DATA:

   Find the value of the test statistic described in your analysis plan. Complete other computations, as required by the plan. In the balance of Unit 4, separate lessons describe each type of analysis and show examples of how to use an algorithm to complete each test.

4. ### INTERPRET RESULTS.

   Apply the decision rule described in the analysis plan. Your research project will be useless without interpreting the results of your research. The analysis plan should include how you will use the data you have gathered to accept or reject the null hypothesis. If the analysis of the data supports the null hypothesis, the null hypothesis is accepted; if the value of the test statistic based on the null hypothesis is unlikely, it is rejected. Your final effort will be to write the report that justifies your interpretation of your data and the conclusion(s) you come to as a result of your study.

UNIT 4A-1: REVIEW - BASIC ESSENTIALS FOR TESTING HYPOTHESES

PURPOSE:

There are some basic principles that underlie the testing of hypotheses. During every Research Project, it is vital to test each hypothesis to determine if there is or is not significance. Hypotheses tests differ depending on which type of data you collect and what you want to achieve. No matter what test you will use, there is a pattern you should follow which applies to **all** hypothesis testing. Of particular importance are the decision rules you must make and the types of errors that may be inherent in the testing process.

Since it is seldom possible to use the population to obtain research data, many of the data tests need to use a sample mean, $\overline{X}$, instead of the population mean, μ, and the sample standard deviation, S_x, instead of the population standard deviation, σ (sigma) There is a problem with using data from a sample: there is an increased likelihood of the sample data not matching the population from which it is drawn. To solve that, researchers use the concept of **standard error**.

OBJECTIVES:

- Define the term "standard error."

- Explain the difference between using standard deviation and standard error.

- Define the term "hypothesized mean."

- Describe the test statistic.

- What are the values that are important when calculating your test statistic?

- Describe the **decision rules** that enable **you** to determine if the data you have is likely to be significant.

- Describe the major types of error that a researcher may make.

- Describe the factors that may affect the power of any analytical test.

- Define "significance."

- Describe the two algorithms for calculating the mean and for calculating the standard deviation.

INTRODUCTION:

A broad generalization of the meaning of the term **null hypothesis** is that the data shows that the actions taken in the study makes no difference or that the results of the various parts of the study are the same.

SAMPLES & POPULATIONS:

In a research study, the researcher must select a sample of the population. The subjects in the sample are selected randomly. Various methods for doing this were discussed in Volume 1, Unit 1B. It is important to realize that if more than one sample were to be selected, randomness would ensure that each sample would be different. There is the possibility that some samples may not truly represent the population, For example, suppose all members of a sample had red hair, but the population members show all colors of hair. This sample would not be a good representation of the population; *it is necessary to ensure that the sample shows the same characteristics as the population*.

When testing data, you may be working with means or proportions. **A proportion is a ratio:** for example, in a town of 600 people, if 250 were females, then the proportion is 250:600. In doing an analysis of the information collected, it is important to distinguish data that applies to the population as a whole and data which applies only to the sample. A simple way of doing this is to use capital letters for population data referred to as **parameters** and lower case letters for sample data referred to as **statistics** as shown in Figure 3.

Population Parameter		Sample Statistic	
N	Number of observations in the population	n	Number of observations in the sample
μ	Population mean	$\bar{x}$	sample estimate of μ
P	Proportion (population)	p	Proportion (sample)
σ	Population standard deviation	s	sample estimate of σ
σ_μ	Standard deviation of μ	$SE_{\bar{x}}$	Standard error of $\bar{x}$

Figure 3: Comparing population parameters & sample statistics.

There are two sets of calculations that can be made when working with means and when working with proportions. One set deals with the standard deviation the other with standard error. Why have both? All too often, the parameters of the population are unknown. As a result, it is not possible to calculate the standard deviation of the sample.

Instead the researcher uses standard error. As you can see from Figure 4, the equations are similar but standard error uses sample statistics in the place of parameters (p instead of P and s instead of μ or **0**). Since the values of these statistics are known or can be calculated, the standard error provides an unbiased estimate of the standard deviation.

Statistic	Standard Deviation	Standard Error
Sample mean: $\bar{X}$	$\sigma_{\bar{X}} = \dfrac{\sigma}{\sqrt{n}}$	$SE_{\bar{X}} = \dfrac{s}{\sqrt{n}}$
difference between sample means: $\bar{X}_1 - \bar{X}_2$	$\sigma_{\bar{X}_1 - \bar{X}_2} = \sqrt{\dfrac{\sigma_1^2}{n_1} + \dfrac{\sigma_2^2}{n_2}}$	$SE_{\bar{X}_1 - \bar{X}_2} = \sqrt{\dfrac{s_1^2}{n_1} + \dfrac{s_2^2}{n_2}}$
sample proportion: p	$\sigma_p = \sqrt{\dfrac{P(1-P)}{n}}$	$SE_p = \sqrt{\dfrac{p(1-p)}{n}}$
difference between sample proportions: $P_1 - P_2$	$\sigma_{P_1 - P_2} = \sqrt{\dfrac{P_1(1-P_1)}{n_1} + \dfrac{P_2(1-P_2)}{n_2}}$	$SE_{P_1 - P_2} = \sqrt{\dfrac{P_1(1-P_1)}{n_1} + \dfrac{P_2(1-P_2)}{n_2}}$

Figure 4: Comparing calculations for the standard deviation and the standard error.

STANDARD ERROR:

Many of the statistical tests need to use a sample mean instead of **μ**, the population mean, and the sample standard deviation instead of **σ** (sigma), the population standard deviation. The problem with doing that is that there is an increased likelihood of the sample data not matching the population from which it is drawn. To solve that, researchers use the concept of **standard erro**r. Since population parameters are usually unknown, statisticians use sample statistics to estimate the population's parameters. The standard error is used to assist in making those estimates more accurate.

Knowing both the standard deviation and the standard error means that you can have a greater confidence in your results. The standard error measures the variability of the statistic you have calculated. Reminder: the term **parameter** refers to a characteristic of the population you are dealing with. A **statistic** is a characteristic of your sample.

The standard error depends on three elements of your population and sample. First are the number of subjects, N, in the population, and second are the number of subjects, n, in the sample. Third is the way the sample is selected, usually by a simple random selection. If all possible samples of **n** objects may be selected, simple random sampling means that, theoretically all samples are equally likely to occur. To ensure that the sample is valid, a level of significance is selected, usually 0.05 or 0.01; O.05 says that your samples will be representative of the population 95 out of 100 times and 0.01 says your samples are representative 99 out of 100 times.

A natural outcome of random selection is that the samples selected, even though they are selected from the same population, may vary. The sample's standard deviation measures that variability. The standard deviation (**σ**) may be calculated depending on what statistic is used. The following formulas work best when the population is at least ten times the size of the sample (e.g., 200 populations means there should be at least 20 in the sample). In addition, the formulas require you to know one or more of the population parameters in order to calculate the standard deviation. Because of the problem in knowing populations parameters, one can instead use the standard error (SE) of the sample statistics which tends to remove bias.

Note that the equations for the standard error are identical to the equations for the standard deviations except for one thing. Standard error equations use statistics (for a sample) where the standard deviation equations use parameters (for the population). Specifically, the standard error equations use p in place of P, and s in place of σ.

A HYPOTHESIZED MEAN:

Because the population mean is seldom known, we may need to **hypothesize the mean**. **What is a *hypothesized mean?*** It is a mean predicted or assumed for the population. How do we arrive at it? We may be able to get one from past results of many tests. For example, IQ has been tested so many times that the average IQ is known to be 100. Of course, if we are working with graduate students, the data may be skewed upwards because it is possible that only people with higher IQs attend graduate school. However for practical purposes, if researching a question concerning IQ, the hypothesized mean can be assumed to be 100.

Another way to arrive at a hypothesized mean is to compare our current group with performance in the past of similar groups. For example, the universities keep records for all students who have graduated. Over a period of time, say the past five years, we can come up with a mean for the university's population and use this in our study. But, we need to be careful in our selection of the population. For example, the population of students that attended the university fifty years ago is likely to be very different from the current population because of historic changes that have occurred in the 50 years. Great-Grandma is likely to be a very different person from her Great-Granddaughter. Make sure that the population you choose for your hypothesized mean is truly similar to the population you are testing.

THE TEST STATISTIC & SIGNIFICANCE:

The purpose of doing any research is to determine if the results of your study are significant. You must create the null hypothesis, essentially saying there is no significant difference in your test results. The alternative hypothesis states what you really want to prove - that there ***are*** significant differences.

You use your data to determine the value of the test statistic as determined by the type of test. The test statistic allows us to test the significance of our findings. Another item that helps us determine significance is the **critical value**, that is, the value that we compare with our **calculated value** to determine if we reject or to fail to reject our test hypothesis. The critical value is determined by the type of test we are using.

There are five tests in this unit: the *z-test,* the *t-test,* either the *Pearson* or the *Spearman tests of correlation* and the *chi-square test.* Each test uses a different formula to calculate your test statistic, the **calculated value**. In addition, a different table for each test is used to determine the **critical value** which you can use to decide whether to reject or fail to reject the null hypothesis.

THE DECISION RULES:

You have a choice of two decision rules. The rules establish how you will analyze your results to determine significance:

1. <u>P-value</u>.

    P refers to the probability of the accuracy of your selected test. If the P-value is less than the significance level, we reject the null hypothesis.

2. <u>Region of acceptance</u>.

    The region of acceptance is a range of values. If the test statistic falls within the region of acceptance, then you can accept the null hypothesis. If the test statistic falls outside the region of acceptance, the null hypothesis is rejected.

DECISION ERRORS:

There are two errors that may be made in any research study:

1. ***Rejecting the null hypothesis when it is true gives a false positive or Type I error.***

2. ***Failing to reject the null hypothesis when it is false gives a false negative or Type II error.***

Be prepared for the possibility of making one or other error in your research.

The probability of making a Type I error is called **alpha** often shown using the Greek lower case letter **α**. This probability (**α**) is referred to as the **significance level**. The most common significance levels are 0.05 meaning your sample has a 95% probability of being similar to the population; or 0.01 a 99% probability.

The probability of making a Type II error is called **beta** often shown using the Greek lower case letter **β**. The probability of *not* making this error is referred to as the **power** of the test. [Note: The source for information on the power of a test is the "Tutorial: Lesson 5. Power" Used with the permission of the author Harvey Berman at the URL <u>http://stattrek.com/Lesson5/Power. aspx?Tutorial=Stat</u>]

FACTORS AFFECTING POWER:

The greater the power of your test the less likely you are to make a Type II error. In using the power to avoid the Type II Error, you determine the **effect size** which is the difference between the true value and the hypothesized value in your study. For example: Let's suppose you are working with I.Q., where the true average value is 100. Suppose the sample you have is expected to be below average I.Q. Your hypothesis might be $H_0 = 90$. So the effect size is 90 - 100 or -10. On the other hand, suppose the sample you have is expected to be above average, your hypothesis might be $H_0 = 110$. The effect size would be 110-100 = 10. There are three factors that affect power.

1. Sample size represented by n: Generally the greater the sample size, the greater the power of the test. This is because the greater the sample size the more closely it approaches the size of the population.

2. Significance level represented by α: Again in general, the higher the significance level the higher the power; a value of 0.01 (99%) for the significance level is higher than a value of 0.05 (95%) which in turn is higher than a significance level of 0.10 (90%). Also, if you **increase** the significance level (for example, to 0.01) you also **reduce** the region of acceptance. As a result you also increase the chance of rejecting the null hypothesis; therefore, you are less likely to accept the null hypothesis when it is false.

3. The "true" value of the hypothesis being tested: The greater the difference between the true value (the parameter of the population) and the hypothesized value (the value for your sample), the greater the power of the test; in other words, the greater the effect size, the greater the power of the test.

ALGORITHMS FOR BASIC TESTS OF INTERVAL DATA:

One method for simplifying the calculations needed for specific analytical tests is to list them as a set of directions called an **algorithm**. As examples of algorithms, the tests of interval data require the calculation of a mean and to use the mean to calculate the standard deviation. You will need to use both in analyzing interval data.

ALGORITHM FOR CALCULATING THE MEAN:

1. **Identify the number of observations (n) in the dataset.**

2. **Find the sum of all measurements (ΣX) in the set. [The symbol Σ represents "the sum of" and the symbol X represents the set of data points to be summed.]**

3. Divide the sum by the number of observations ($\Sigma X/n$) to get the mean $\overline{X}$.

ALGORITHM FOR CALCULATING THE VARIANCE & STANDARD DEVIATION:

1. **Calculate the mean ($\overline{X}$) using the algorithm above.**

2. **Calculate the deviation from the mean for each and every measurement in the dataset [X - 0]**

3. **Square each result in step 2 to get each squared deviation.**

4. **Sum the squares of all deviations [$\Sigma(X - \overline{X})2$]**

5. **Divide that sum by n-1 to get the variance, [$\Sigma(X - \overline{X})2/(n-1)$]**

6. **Take the square root of the variance to get the standard deviation, [$\sqrt{\{(X - 0)2/(n-1)\}}$]**

UNIT 4A-1: ASSIGNMENT.

1. **Describe how the four steps in the process of testing hypotheses will be of value in your research project.**

2. **Describe the decision rules that enable you to accept or reject the null hypothesis.**

3. <u>**Recommended:**</u> **Memorize the two basic algorithms.**

UNIT 4A-2. ANALYSIS OF HYPOTHESES

PURPOSE:

The basic hypothesis tests deal with specific types of data. Both the z-test and the t-test are used on interval data. There are two correlation tests, one is the Pearson Product Moment test which is used for interval data and the other is the Spearman Rank Order test which is used for ordinal data. Finally, there is the Chi-Square test which is used for nominal data.

One important aspect of our analyses is that we can use a specific test and find the **test statistic**. The test statistic allows us to test the significance of our findings. Another item that helps us determine significance is the **critical value**, that is, the value that we compare with our **calculated value** to determine if we reject or fail to reject our null hypothesis.

It will be important to understand the basics of data analysis so you can make sure that you can talk intelligently to the statistic department graduate student who may assist you. Even if you cannot remember the analytical tests themselves, you need to understand the purpose of each in order to analyze your dissertation or thesis data.

OBJECTIVES:

- List each of the tests that can be used to analyze your data and what type of data each one applies to..
- Define "hypothesized mean".
- Describe how you can determine whether to reject the null hypothesis or fail to reject.

INTRODUCTION:

There are specific tests for each type of data. When you are determining your analysis plan, you must decide which test or tests are needed for the data you have collected. In this first study it was recommended you collect all types of data (interval, ordinal and nominal) so you can get the experience of doing each of the tests.

WHAT CAN WE ANALYZE?:

The ideal way to conduct research is to obtain data from the whole of the population in which we are interested. For interval data, this enables us, for example, to make a comparison between the population mean, μ or **sigma**, and the sample mean. However, in most cases, trying to obtain data from the whole population would be impractical and, in real life, we rarely know exactly what μ is for the whole population. There are methods where we can substitute the sample mean for μ. **To do this, we must find a way to determine if our sample is truly representative of the population.**

One of the simplest tests we can do is analyze means. In addition to analyzing the means of our datasets, we can analyze proportions. Analyzing a proportion allows us to compare a set of data with n, the total number of data items we have. A proportion may be written as a ratio, such as 5:10 meaning that we are looking at five of ten items. It can also be written as a fraction, one half or ½, or as a percentage, 50%.

We can also examine relationships, such as the relationship between scores for reading and spelling, to determine if people who read well also spell well. Another example: there may be a relationship between the scores in the midterm exam and the final exam. One might expect those who did will in the midterm to also do well in the final, and those who did poorly in one would do poorly in the other.

During the analysis process, one step we would need to take is to determine if one set of data is dependent on the other. For example, if the comparison between the midterm and the final exam uses the same set of students, then the relationship is dependent. If we compared the final exam in the same subject for one set of students with the same final exam of a different set of students, the two sets are independent even though the exams are the same. We may compare the rankings of the same set of students in two different subjects or we may compare the rankings of different students in the same subject. No matter which method we use to compare such data, we may also wish to know if there is a good fit between compared sets of data. Also we may think our two (or more) sets of data are independent, but it may be wise to test for independence. And since we may have used a population parameter in our analysis, we may want to test how well the sample variance compares with the population variance. The tests to be considered in this unit deal with these kinds of questions.

ANALYSIS OF MEANS OR PROPORTIONS (INTERVAL DATA):

There are several tests involving means and proportions of a dataset. The first of these, the **z-test**, is usually used with a **large sample**. The z-test can be used for testing one or two means or one or two proportions. Proportions can be written as a ratio such as 5:10, or a fraction, ½, or as a percentage, 50%. There are three ways that a t-test can be used.

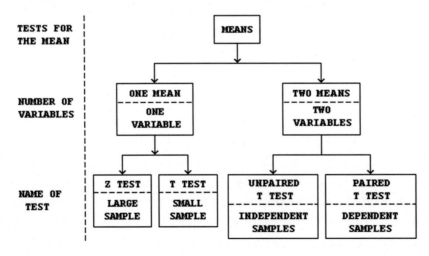

Figure 5: Test of means (interval data).

The first way is where a single mean is involved (i.e., a single variable or dataset). Second is where two means are involved (i.e., two variables) with the two datasets having **unpaired** and **independent samples** or **paired** and **dependent samples**. An independent sample occurs when the subjects involved in one of the datasets are totally different from the subjects in the other. The dependent sample occurs when the subject are the same with two different datasets.

For example, the sample may be data from the sixth grade from two different schools; the subjects, being totally different sets from the two schools, are independent. Two different sections with no overlap of students at university would also be independent.

The third test of means is a **paired test** with **dependent samples**. For example, the same set of students tested at two different times such as at the beginning of the school year and again at the end of the year; or in two different subjects such as Math and English or Science and Creative Writing.

Each of the tests involving means will be discussed in detail in the lessons titled **4B: USING THE Z-TEST, with** the new concept for z-tests being *Standard Error*; and **4C: BASIC PRINCIPLES FOR THE T-TEST,** with the new concept for t-tests being *Degrees of Freedom*.

CORRELATIONS ANALYSIS:

"Correlation" refers to a comparison between two sets of data to determine if there is a relationship between the two sets. There are two Correlation tests. The first is the *Pearson Product Moment Test* which tests the correlation (relationship) between two variables of interval data. The second is the *Spearman Rank Correlation Test* which tests ordinal data to find the correlation between the rankings in two sets of variables.

The direct formula is an alternative method of calculating the correlation coefficient, **r,** when the data is simple (the data values are less than 10). We will use both formulas. A new concept is **estimated means.**

When dealing with ranked data, sometimes the data is already ranked. In other datasets, the rank is not given and must be added.

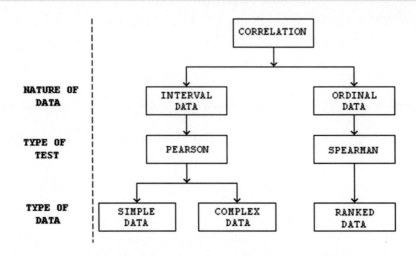

Figure 6: Two correlation tests; Pearson's used with interval data and Spearman's used with ordinal data.

USE OF THE CHI-SQUARE TESTS:

When the data is nominal or categorical the Chi Square test can be used. The Chi symbol is the Greek letter χ.

The first test is the Chi-square goodness-of-fit where there is only one variable. This test requires you to have a hypothesized (expected) distribution of data and the test determines if the collected data fits the hypothesized distribution.

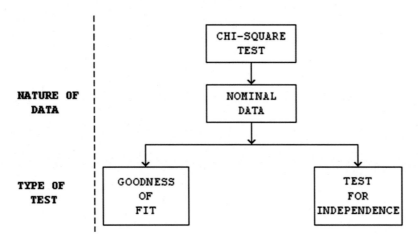

Figure 7: Chi-Square tests for nominal data.

The second test is the Chi-square test for independence where there are two categorical variables and you plan to test the amount of association between the two variables... In addition, there is a comparison made between **observed frequencies,** the data you gather, and **expected frequencies**, the data the test suggests you **should** get.

The third test enables you to test for the population variance and to compare it with the sample variance.

UNIT 4A-2: ASSIGNMENT.

1. **Define the term "standard Error" and describe why it is so important.**

2. **Describe the differences between a one-tailed and two-tailed test including the identifying hypotheses.**

UNIT 4A-3: IMPORTANT POINTS ABOUT ONE- & TWO-TAILED TESTS

PURPOSE:

Now that you are ready to start working on the various tests, it is important to understand, in terms of hypotheses, the difference between one- and two-tailed tests. This is of particular value when you hypothesis requires either one or both.

OBJECTIVE:

- Describe how the null and alternative hypotheses determine what type of test (one- or two-tailed) you will use.

- Describe the differences between a one-tailed and two-tailed test including the identifying hypotheses.

- Look at the hypotheses you have developed and determine if the information in this lesson has an impact on your hypotheses.

- Describe how the way the null hypothesis is written determines the region or regions of acceptance and rejection.

- Describe what is meant by "variability between groups."

TESTS FOR SPECIFIC TYPES OF DATA

Researchers work with **nominal** data i.e. categorical data, or **ordinal** i.e. data that is ranked, or **interval** data i.e. data that has numerical values. Since each type of data is different, it follows that different test must be used on different types of data.

There are five hypothesis tests that may be used.

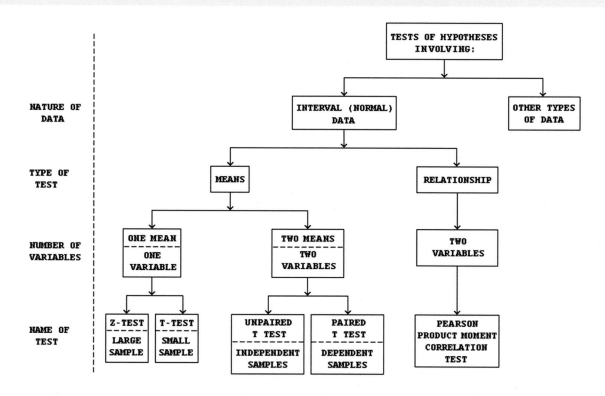

Figure 8: the main analysis paths for interval data.

Figure 8 shows the main paths for using the *z-test;* the *t-test;* and the ***Pearson product moment correlation test***. All these tests are used on interval data. The two other types of data considered in this text book, are ordinal and nominal. Figure 6 shows the main paths for these tests.

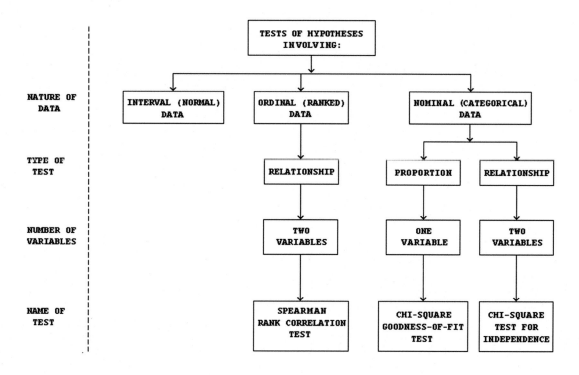

Figure 9: Data tests for ordinal and nominal data.

ONE-TAILED & TWO-TAILED T-TESTS:

The null hypothesis tells us if we are dealing with a one- or two-tailed test. Although one tends to consider one-tailed test; the simpler explanation deals with a two-tailed test.

For example, if H_0 contains only the equals sign (=), the alternative hypothesis H_a contains the "not equals" sign (≠). "Not equals to" means that any value other than the one specified by the null hypothesis determines whether we can reject H_0. "Not equal to" refers to any value less than or any value more than the value shown in H_0, in other words, again there are two tails.

In Figure 9a, the red line represents H_0. Even though the red line represents only a single value, it divides each curve into two segments each of which has a separate tail. The green areas on either side of the red line represent H_a that is all values not equal to **the value of H_0**. In other words, H_a has a range of values.

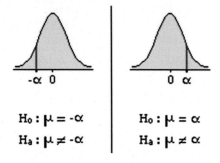

Figure 10a: The two-tail test, with the null hypothesis having a single value.

Figure 10a shows the effect of using "greater than or equal to" or "less than or equal to" for H_0 and "not equal to" for H_a.

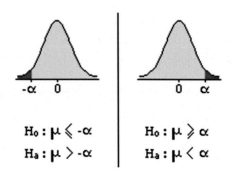

Figure 10b: One-tail tests with the null hypothesis having a range of values.

In Figure 10b, the null hypothesis contains both the equality and inequality, in this case either "greater than or equal to α" or "less than or equal to -α." The use of ≤-α means: every point on the curve from the end of the curve to −α. The use of ≥α means: every point on the curve from α to the end of the curve. In each case, only one end of the curve is involved hence only a single tail.

Since the two-tailed graphs and the one-tailed graphs are different, the full t-table has two different sets of table values. One when there is a single tail, one where there are two tails.

t Table

one-tail	0.50	0.25	0.20	0.15	0.10	0.05	0.025	0.01	0.005	0.001	0.0005
two-tails	1.00	0.50	0.40	0.30	0.20	0.10	0.05	0.02	0.01	0.002	0.001
df 1	0.000	1.000	1.376	1.963	3.078	6.314	12.71	31.82	63.66	318.31	636.62
2	0.000	0.816	1.061	1.386	1.886	2.920	4.303	6.965	9.925	22.327	31.599
3	0.000	0.765	0.978	1.250	1.638	2.353	3.182	4.541	5.841	10.215	12.924
4	0.000	0.741	0.941	1.190	1.533	2.132	2.776	3.747	4.604	7.173	8.610
5	0.000	0.727	0.920	1.156	1.476	2.015	2.571	3.365	4.032	5.893	6.869
6	0.000	0.718	0.906	1.134	1.440	1.943	2.447	3.143	3.707	5.208	5.959
7	0.000	0.711	0.896	1.119	1.415	1.895	2.365	2.998	3.499	4.785	5.408
8	0.000	0.706	0.889	1.108	1.397	1.860	2.306	2.896	3.355	4.501	5.041
9	0.000	0.703	0.883	1.100	1.383	1.833	2.262	2.821	3.250	4.297	4.781
10	0.000	0.700	0.879	1.093	1.372	1.812	2.228	2.764	3.169	4.144	4.587
11	0.000	0.697	0.876	1.088	1.363	1.796	2.201	2.718	3.106	4.025	4.437
12	0.000	0.695	0.873	1.083	1.356	1.782	2.179	2.681	3.055	3.930	4.318
13	0.000	0.694	0.870	1.079	1.350	1.771	2.160	2.650	3.012	3.852	4.221
14	0.000	0.692	0.868	1.076	1.345	1.761	2.145	2.624	2.977	3.787	4.140
15	0.000	0.691	0.866	1.074	1.341	1.753	2.131	2.602	2.947	3.733	4.073
16	0.000	0.690	0.865	1.071	1.337	1.746	2.120	2.583	2.921	3.686	4.015
17	0.000	0.689	0.863	1.069	1.333	1.740	2.110	2.567	2.898	3.646	3.965
18	0.000	0.688	0.862	1.067	1.330	1.734	2.101	2.552	2.878	3.610	3.922
19	0.000	0.688	0.861	1.066	1.328	1.729	2.098	2.539	2.861	3.579	3.883
20	0.000	0.687	0.860	1.064	1.325	1.725	2.086	2.528	2.845	3.552	3.850
21	0.000	0.686	0.859	1.063	1.323	1.721	2.080	2.518	2.831	3.527	3.819
22	0.000	0.686	0.858	1.061	1.321	1.717	2.074	2.508	2.819	3.505	3.792
23	0.000	0.685	0.858	1.060	1.319	1.714	2.069	2.500	2.807	3.485	3.768
24	0.000	0.685	0.857	1.059	1.318	1.711	2.064	2.492	2.797	3.467	3.745
25	0.000	0.684	0.856	1.058	1.316	1.708	2.060	2.485	2.787	3.450	3.725
26	0.000	0.684	0.856	1.053	1.315	1.706	2.056	2.479	2.779	3.435	3.707
27	0.000	0.684	0.855	1.057	1.314	1.703	2.052	2.473	2.771	3.421	3.690
28	0.000	0.683	0.855	1.056	1.313	1.701	2.043	2.467	2.763	3.408	3.674
29	0.000	0.683	0.854	1.055	1.311	1.699	2.045	2.462	2.756	3.396	3.659
30	0.000	0.683	0.854	1.055	1.310	1.697	2.042	2.457	2.750	3.385	3.646
40	0.000	0.681	0.851	1.050	1.303	1.684	2.021	2.423	2.704	3.307	3.551
60	0.000	0.679	0.848	1.045	1.296	1.671	2.000	2.390	2.660	3.232	3.460
80	0.000	0.678	0.846	1.043	1.292	1.664	1.990	2.374	2.639	3.195	3.416
100	0.000	0.677	0.845	1.042	1.290	1.660	1.984	2.364	2.626	3.174	3.390
1000	0.000	0.675	0.842	1.037	1.282	1.646	1.962	2.330	2.531	3.093	3.300
	0%	50%	60%	70%	80%	90%	95%	98%	99%	99.8%	99.9%

Confidence Level

Figure 11a: The t-table with values for both one- and two-tailed tests.

Let's suppose that in Figure 10b, that H_0: $\mu \geq 24$. This means we have a one-tailed test with H_a being all values **less than** 24. In the t-table, look along the line with 24 **degrees of freedom** and down from the **one-tail line** at **0.01 significance level**.

t Table

df	one-tail 0.50 two-tails 1.00	0.25 0.50	0.20 0.40	0.15 0.30	0.10 0.20	0.05 0.10	0.025 0.05	(0.01) 0.02	0.005 (0.01)	0.001 0.002	0.0005 0.001
1	0.000	1.000	1.376	1.963	3.078	6.314	12.71	31.82	63.66	318.31	636.62
2	0.000	0.816	1.061	1.386	1.886	2.920	4.303	6.965	9.925	22.327	31.599
3	0.000	0.765	0.978	1.250	1.638	2.353	3.182	4.541	5.841	10.215	12.924
4	0.000	0.741	0.941	1.190	1.533	2.132	2.776	3.747	4.604	7.173	8.610
5	0.000	0.727	0.920	1.156	1.476	2.015	2.571	3.365	4.032	5.893	6.869
6	0.000	0.718	0.906	1.134	1.440	1.943	2.447	3.143	3.707	5.208	5.959
7	0.000	0.711	0.896	1.119	1.415	1.895	2.365	2.998	3.499	4.785	5.408
8	0.000	0.706	0.889	1.108	1.397	1.860	2.306	2.896	3.355	4.501	5.041
9	0.000	0.703	0.883	1.100	1.383	1.833	2.262	2.821	3.250	4.297	4.781
10	0.000	0.700	0.879	1.093	1.372	1.812	2.228	2.764	3.169	4.144	4.587
11	0.000	0.697	0.876	1.088	1.363	1.796	2.201	2.718	3.106	4.025	4.437
12	0.000	0.695	0.873	1.083	1.356	1.782	2.179	2.681	3.055	3.930	4.318
13	0.000	0.694	0.870	1.079	1.350	1.771	2.160	2.650	3.012	3.852	4.221
14	0.000	0.692	0.868	1.076	1.345	1.761	2.145	2.624	2.977	3.787	4.140
15	0.000	0.691	0.866	1.074	1.341	1.753	2.131	2.602	2.947	3.733	4.073
16	0.000	0.690	0.865	1.071	1.337	1.746	2.120	2.583	2.921	3.686	4.015
17	0.000	0.689	0.863	1.069	1.333	1.740	2.110	2.567	2.898	3.646	3.965
18	0.000	0.688	0.862	1.067	1.330	1.734	2.101	2.552	2.878	3.610	3.922
19	0.000	0.688	0.861	1.066	1.328	1.729	2.098	2.539	2.861	3.579	3.883
20	0.000	0.687	0.860	1.064	1.325	1.725	2.086	2.528	2.845	3.552	3.850
21	0.000	0.686	0.859	1.063	1.323	1.721	2.080	2.518	2.831	3.527	3.819
22	0.000	0.686	0.858	1.061	1.321	1.717	2.074	2.508	2.819	3.505	3.792
23	0.000	0.685	0.858	1.060	1.319	1.714	2.069	2.500	2.807	3.485	3.768
(24)	0.000	0.685	0.857	1.059	1.318	1.711	2.064	(2.492)	(2.797)	3.467	3.745
25	0.000	0.684	0.856	1.058	1.316	1.708	2.060	2.485	2.787	3.450	3.725
26	0.000	0.684	0.856	1.053	1.315	1.706	2.056	2.479	2.779	3.435	3.707
27	0.000	0.684	0.855	1.057	1.314	1.703	2.052	2.473	2.771	3.421	3.690
28	0.000	0.683	0.855	1.056	1.313	1.701	2.043	2.467	2.763	3.408	3.674
29	0.000	0.683	0.854	1.055	1.311	1.699	2.045	2.462	2.756	3.396	3.659
30	0.000	0.683	0.854	1.055	1.310	1.697	2.042	2.457	2.750	3.385	3.646
40	0.000	0.681	0.851	1.050	1.303	1.684	2.021	2.423	2.704	3.307	3.551
60	0.000	0.679	0.848	1.045	1.296	1.671	2.000	2.390	2.660	3.232	3.460
80	0.000	0.678	0.846	1.043	1.292	1.664	1.990	2.374	2.639	3.195	3.416
100	0.000	0.677	0.845	1.042	1.290	1.660	1.984	2.364	2.626	3.174	3.390
1000	0.000	0.675	0.842	1.037	1.282	1.646	1.962	2.330	2.531	3.093	3.300
	0%	50%	60%	70%	80%	90%	95%	98%	99%	99.8%	99.9%

Confidence Level

Figure 11b: The t-table showing the one-tail result (red) and two-tail result (green) for 24 df.

One-tail: The column meets the row for 24 df at the value t = 2.492. This value is the same whether the tail is on the left **or** on the right (either curve in Figure 10a). If we are considering the two-tailed test (Figure 10b), the null hypothesis would be H_0: $\mu = 24$. Then H_a is any value **not equal to 24.** That means it can be either any value greater than 24 or it can be any value less than 24 (two tails). Reading the value from the chart (green), t = 2.797.

The same t-table can be used for either one- or two-tails, but notice that for the two tails (see the second title row) the significance levels are shifted one column to the right.

DETERMINING THE NULL AND ALTERNATIVE HYPOTHESES:

For the previous example, $H_0: \mu \geq 24$ means we have a one-tailed test with H_a being all values less than 24. Look along the line with **9 degrees of freedom** and down from the **one-tail line** at **0.05 significance level**. The column meets the row for 9 degrees of freedom at the value 1.833. This value is correct whether the tail is on the left **or** on the right of the t-value (either curve in Figures 10a and 10b). If we are considering the two-tailed test, the null hypothesis in the previous example **would have to be** $H_0: \mu = 24$. Then H_a is any value **not equal to 24.** That means it can be either any value greater than or any value less than 24. NOTE: *Using either $\leq$ or $\geq$ gives a one-tailed test and using = gives a two-tailed test.*

Let's look again at the earlier example: Five years ago, a college became a university and now the administration wishes to know if the freshman intake has remained the same as freshmen intake when the institution was a college. The university has collected data for the years it was a college and the years since it has become a university.

Suppose the **null hypothesis** is that the status of the institution does not affect the student intake, then $H_0: \overline{X} = \mu$, where $\overline{X}$ is the average of the five years since becoming a university and μ is the **average prior to becoming a university**. This is a two-tail test. The **alternative hypothesis** would be that the change in status of the institution does affect the student intake, therefore, $H_a: \overline{X} \neq \mu$. The hypothesized mean would be μ or the mean for the years the institution was a college which might be limited to the last five years the institution was a college. The observational mean could be either a mean for the latest year as a university or a mean for the five years since the conversion. In either case, the observational mean is being compared to the hypothesized mean based on prior records. This is a two-tailed test and any significant variance of the mean, $\overline{X}$, from the hypothesized mean, μ, would result in the rejection of the null hypothesis.

Suppose, however, the administration feels that the current intake must be of better quality, as reflected by ACT scores, than those when the institution was just a college. Then Ha: $\overline{X} > \mu$ with $\overline{X}$ being the mean ACT scores for current freshmen and μ being the mean ACT scores for the intake of students for the last year when the institution was a college. We start with the alternative hypothesis because the expectation is for higher average ACT scores. The > sign is an inequality hence it defines the alternative hypothesis $H_a: \overline{X} > \mu$. H_0 and H_a are complements of each other, therefore the area of the curve defined by the null hypothesis is the complement, defined by $\leq$, of the area defined by the alternative hypothesis. Therefore, the null hypothesis is H0: $0 \leq \mu$. "Less than or equal to" includes the possibility that there has been no change in ACT Scores. It also (perish the thought) includes the possibility that the quality of freshmen has actually deteriorated.

This null hypothesis of $H_0: \overline{X} \leq \mu$ means this is a one-tail test. The alternative hypothesis of $\overline{X} > \mu$ means that this is a right-tailed test. An alternative hypothesis of $H_0: \overline{X} \geq \mu$ and $H_a: \overline{X} < \mu$ would mean a left-tailed test.

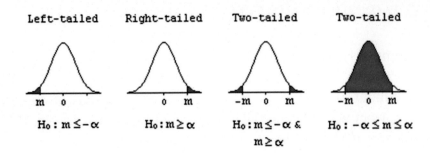

Figure 12: Null hypotheses for the various types of tailed curves.

REGION OF ACCEPTANCE:

Now that you are ready to start working on the various tests, it is a good time to review some of the information in the *Unit 2B. Research Hypotheses* from a slightly different viewpoint. It is important to understand, in terms of hypotheses, the difference between one- and two-tailed tests. This is of particular value when your data requires a one- or a two-tailed test.

The null hypothesis tells us if we are dealing with a one- or two- tailed test. For example, if **H₀** contains only the equals sign (=), the alternative hypothesis **Hₐ** contains the "not equals" sign (≠). "Not equals to" means that any value other than the one specified by the null hypothesis determines whether we can reject H₀. "Not equals to" refers to any value less than or any value more than the value shown in H₀, in other words, two tails.

Suppose we wish to test the population measure M and the mean of the population is μ. Then the null and alternate hypotheses are:

$$H_0: \mu = M \qquad \text{Region of acceptance of null hypothesis.}$$
$$H_a: \mu \neq M \qquad \text{Region of rejection of null hypothesis.}$$

The symbol ":" is read as "such that." In the above hypotheses, the two are read as: The null hypothesis such that μ is equal to the selected value represented by M. The alternative hypothesis such that μ is not equal to the selected value of M. The symbol ≠ means "not equal to"; some researchers use the following for the same meaning: $H_a: \mu <> M$ where <> means that μ is either less than M or more than M. In other words μ is on either side of M, but is not equal to M itself.

Using the letter M is arbitrary and it is used to represent your data which may be a single value or a range of values. Any letter may be used from the alphabet, such as M or m, n, p, or any letter of the Greek alphabet, such as α or β (Greek letter for a and b). For example, in the following graphic, α is used.

Set	Null hypothesis	Alternative hypothesis
1	$\mu = \alpha$	$\mu \neq \alpha$
2	$\mu \geq \alpha$	$\mu < \alpha$
3	$\mu \leq \alpha$	$\mu > \alpha$

Figure 13: Comparing the null and alternative hypotheses.

As shown in Figure 13, there are three basic possibilities for the null hypothesis. The population mean is either equal to a certain value such as α; greater than or equal to α; or less than or equal to α. Note that in each of the three possibilities, there is an equality. With the alternative hypothesis there is no equality. In the first case, the population mean is **not equal to** α; in the second it is **less than** α; and in the third it is **greater than** α. Note also that between them, the null and the alternative hypothesis on the same line make one whole mathematically. In fact the alternative hypothesis is the complement of (completes) the null hypothesis.

When working with interval data, the data can be visualized as a curve. A normal curve is also called a bell curve.

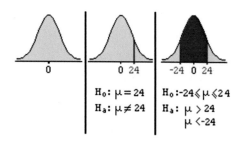

Figure 14: Left - a normal or Bell curve; Middle - H_0 has a single value; Right – H_0 has a range of values.

For the selected values in Figure 15, the region(s) of acceptance of H_0 are marked in red and the region(s) of rejection of H_0 are marked in green. In other words, in the middle illustration, H_0 is accepted if and only if H_0 = 24 (red area), and rejected if it has any other value (green areas). In the illustration on the right, H_0 is accepted if and only if its value lies between -24 and +24 (red area) and rejected if its value lies outside that range (green areas).

Set	Null hypothesis	Alternative hypothesis	Number of tails
1	$\mu = \alpha$	$\mu \neq \alpha$	2
2	$\mu \geq \alpha$	$\mu < \alpha$	1
3	$\mu \leq \alpha$	$\mu > \alpha$	1

Figure 15: The number of tails shown for each set.

The only set that has two tails is the one where there is an equality without a greater than or less than symbol. The value of α may be a single value or a range of values. The other two sets (2 & 3) are graphed as shown in Figure 15.

Notice that in these two cases, there are both equality and inequality symbols and α has a range of values. Note also, the red area under each curve is the region of acceptance of H_0; the green area under each curve is the region of rejection of H_0 and the acceptance of H_a.

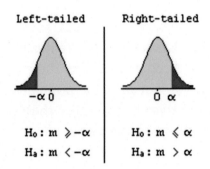

Figure 16: The data determine curves with only one tail.

t Table

one-tail	0.50	0.25	0.20	0.15	0.10	0.05	0.025	0.01	0.005	0.001	0.0005
two-tails	1.00	0.50	0.40	0.30	0.20	0.10	0.05	0.02	0.01	0.002	0.001
df 1	0.000	1.000	1.376	1.963	3.078	6.314	12.71	31.82	63.66	318.31	636.62
2	0.000	0.816	1.061	1.386	1.886	2.920	4.303	6.965	9.925	22.327	31.599
3	0.000	0.765	0.978	1.250	1.638	2.353	3.182	4.541	5.841	10.215	12.924
4	0.000	0.741	0.941	1.190	1.533	2.132	2.776	3.747	4.604	7.173	8.610
5	0.000	0.727	0.920	1.156	1.476	2.015	2.571	3.365	4.032	5.893	6.869
6	0.000	0.718	0.906	1.134	1.440	1.943	2.447	3.143	3.707	5.208	5.959
7	0.000	0.711	0.896	1.119	1.415	1.895	2.365	2.998	3.499	4.785	5.408
8	0.000	0.706	0.889	1.108	1.397	1.860	2.306	2.896	3.355	4.501	5.041
9	0.000	0.703	0.883	1.100	1.383	1.833	2.262	2.821	3.250	4.297	4.781
10	0.000	0.700	0.879	1.093	1.372	1.812	2.228	2.764	3.169	4.144	4.587
11	0.000	0.697	0.876	1.088	1.363	1.796	2.201	2.718	3.106	4.025	4.437
12	0.000	0.695	0.873	1.083	1.356	1.782	2.179	2.681	3.055	3.930	4.318
13	0.000	0.694	0.870	1.079	1.350	1.771	2.160	2.650	3.012	3.852	4.221
14	0.000	0.692	0.868	1.076	1.345	1.761	2.145	2.624	2.977	3.787	4.140
15	0.000	0.691	0.866	1.074	1.341	1.753	2.131	2.602	2.947	3.733	4.073
16	0.000	0.690	0.865	1.071	1.337	1.746	2.120	2.583	2.921	3.686	4.015
17	0.000	0.689	0.863	1.069	1.333	1.740	2.110	2.567	2.898	3.646	3.965
18	0.000	0.688	0.862	1.067	1.330	1.734	2.101	2.552	2.878	3.610	3.922
19	0.000	0.688	0.861	1.066	1.328	1.729	2.098	2.539	2.861	3.579	3.883
20	0.000	0.687	0.860	1.064	1.325	1.725	2.086	2.528	2.845	3.552	3.850
21	0.000	0.686	0.859	1.063	1.323	1.721	2.080	2.518	2.831	3.527	3.819
22	0.000	0.686	0.858	1.061	1.321	1.717	2.074	2.508	2.819	3.505	3.792
23	0.000	0.685	0.858	1.060	1.319	1.714	2.069	2.500	2.807	3.485	3.768
24	0.000	0.685	0.857	1.059	1.318	1.711	2.064	2.492	2.797	3.467	3.745
25	0.000	0.684	0.856	1.058	1.316	1.708	2.060	2.485	2.787	3.450	3.725
26	0.000	0.684	0.856	1.053	1.315	1.706	2.056	2.479	2.779	3.435	3.707
27	0.000	0.684	0.855	1.057	1.314	1.703	2.052	2.473	2.771	3.421	3.690
28	0.000	0.683	0.855	1.056	1.313	1.701	2.043	2.467	2.763	3.408	3.674
29	0.000	0.683	0.854	1.055	1.311	1.699	2.045	2.462	2.756	3.396	3.659
30	0.000	0.683	0.854	1.055	1.310	1.697	2.042	2.457	2.750	3.385	3.646
40	0.000	0.681	0.851	1.050	1.303	1.684	2.021	2.423	2.704	3.307	3.551
60	0.000	0.679	0.848	1.045	1.296	1.671	2.000	2.390	2.660	3.232	3.460
80	0.000	0.678	0.846	1.043	1.292	1.664	1.990	2.374	2.639	3.195	3.416
100	0.000	0.677	0.845	1.042	1.290	1.660	1.984	2.364	2.626	3.174	3.390
1000	0.000	0.675	0.842	1.037	1.282	1.646	1.962	2.330	2.531	3.093	3.300
	0%	50%	60%	70%	80%	90%	95%	98%	99%	99.8%	99.9%
					Confidence Level						

Figure 17: Note the differences in the significance levels for one- and two-tail hypotheses. Also note the percentages for each level (at the bottom of the chart).

VARIABILITY BETWEEN GROUPS:

When working with testing hypotheses, it is important to realize that in making the comparison of the means, we must not just look at given values and make our inferences that the means are or are not equal directly from these values. The following example clarifies this point.

Suppose we are interested in comparing the mean cost of books during the last ten years, published by two different publishers (X and Y). Assume that we take a random sample of 100 books published by each publisher, and find that the mean cost of books published by X is $40, and the mean cost of books published by Y is $50. Can we draw the conclusion that Y publishes more expensive books than X?

The answer is no! Since the selection was made at random, it is possible that we missed getting any of X's more expensive books, with the result the data produced a mean not consistent with the true value of the mean of population **X**.

The illustration below shows two different sets of data that could produce the means of $40, and $50. In the second of the two graphs, the population of **X** is **right-tailed**.

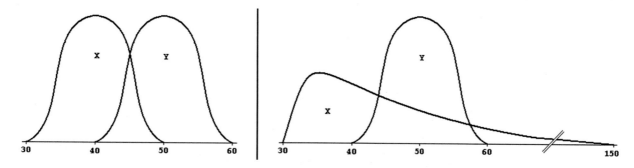

Figure 19: While X publishes more expensive books, the means show that Y publishes the greater number of expensive books.

The left illustration in Figure 19 shows the price ranges are such that the means clearly reflect whose publications are more expensive. On the other hand, Figure 2 shows the situation where X publishes a few very expensive books, but, because the majority of X's books are in the lower price range ($30 to $50), the mean for X is less than that for Y, whose price range is between $40 and $60. Also this data doesn't even take into consideration the fact that the cost of text books is spiraling upward rapidly. However, in spite of these considerations, it is still possible to compare means.

The purpose of doing any research is to determine if the results of your study are significant. You must create the null hypothesis, essentially saying there is no significant difference in your test results. The alternative hypothesis states what you really want to prove - that there **are** significant differences. You use your data to determine the value of the test statistic as determined by the type of test.

The tests in this unit are the **z-test,** the **t-test,** either the **Pearson or the Spearman tests of correlation** and the **chi-square test.** Each test uses a different formula to calculate your test statistic, the **calculated value**. In addition, a different table for each test is used to determine the **critical value** which you can use to decide whether to reject or fail to reject the null hypothesis.

UNIT 4A-3. ASSIGNMENT:

1. **Describe the differences between a one-tailed and two-tailed test including the identifying hypotheses.**

UNIT 4B: THE Z-TEST & Z-SCORE FOR ANALYZING INTERVAL DATA

PURPOSE:

A **z-score** is also known as a **standard score.** Its main use is with large samples (30 or more) when a standardized score is needed; standardizing the score produces a normal distribution. If your dataset cannot be graphed as a normal curve, by calculating the z-scores, the resulting curve shows a normal distribution.

The purpose of this unit is to show you how to convert a raw score into a z-score with a given sample **mean** and a given **standard deviation,** then to determine if the result enables you to reject the null hypothesis, H_0. Two methods can be used (P-value = Probability value):

1. Look up the table to find the **P-value** for the **calculated z-score**, for a predetermined significance level, usually 0.01 or 0.05. *If the P-value is less than the significance level, reject H_0.*

2. Determine the **critical value** of the z-score based on the level of significance. *If the calculated z-value is greater than the critical value, then reject H_0.*

OBJECTIVES:

- Describe how the mean, standard deviation and the set of observations enable you to calculate the set of z-scores.

- Given a sample mean and the sample's standard deviation, calculate a set of z-scores and use a number line to illustrate it.

- Describe the use of a z-score with a single mean, two means, a single proportion and two proportions.

- Given a single variable, determine the P-value for the rejection region in a one-tailed test and decide if H_0 can be rejected.

- Given two variables, determine the P-value for the rejection region in a two-tailed test and decide if H_0 can be rejected.

- If your data gives you a single mean, use the z-test to convert the raw scores to z-scores, and illustrate the result.

- Given your own interval data, calculate the set of z-scores and illustrate the result and determine if H_0 can be rejected.

INTRODUCTION:

The z-test is used only with interval data and the sample should be large. The test is determined by whether the measure used in the calculations is one or two means, or one or two proportions.   The z-test is used to test a single mean, two means, a single proportion, or two proportions. **A proportion is a ratio.** For example, in a town of 600 people, if 280 were females. Then the proportion is 280:600.

THE FOUR BASIC Z-TESTS:

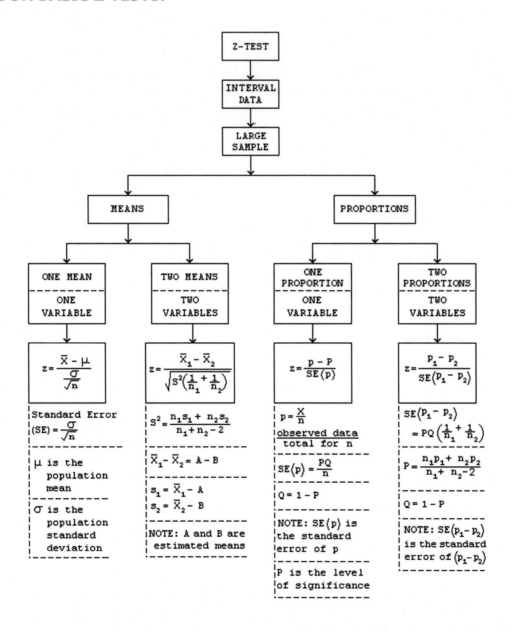

Figure 20: The flow chart shows the four paths for the z-test.

There are four "paths" for conducting z-tests: the first two paths use means; the second two paths use ratios or proportions.

When working with means, you can test a single mean which indicates that your dataset has only one variable. Or you can test two means which indicates that your dataset has two variables. When working with proportions, again you can have a single proportion which indicates one variable in your dataset, or two proportions which indicates two variables.

A different formula is used for each of these paths as shown in Figure 20. The information inside the dashed lines is explanatory information.

THE Z-SCORE IS DERIVED FROM A RAW SCORE:

The z-test converts a **raw score** into a **z-score**, which in terms of statistical analysis is a standardized score. A normal standardized distribution has a mean of zero and a standard deviation of 1 (one)[2]. [Note the second footer number is here]

The majority of z-scores lie between -3 and +3. By converting raw scores into z-scores we can describe raw scores in terms of *standard deviations from the mean.* Therefore a z-value is a measure of **position relative to the mean**.

There are two population parameters used in calculating a z-score: μ, the Greek lowercase letter 'm' or 'mu', pronounced mew, representing the population mean or a hypothesized mean and the population standard deviation, σ, the lowercase Greek letter 's' or 'sigma.'

Since the population mean is rarely known, sigma must also be unknown. So the calculation substitutes the sample mean ($\overline{X}$) for the population mean and the sample standard deviation S_x for sigma (σ). Because of the substitution, the z-scores are not as useful as some of the other statistical measures. But it does give a chance to compare an individual's z-score relative to z-scores of the rest of the students.

NEED FOR A STANDARD SCORE:

A single score in isolation does not give us much information. For example, some time ago a person scored 55 in a geometry test. There is no way to tell how well this person did unless we can make some kind of comparison. If we know the possible total was 100, we can say she did not do very well. But how does her score compare with the rest of the people in the class? Suppose we are told that she got the highest score. This indicates that either the members of the class were not very good at geometry; or the test was too difficult; or the teacher didn't teach very well. Having the additional information enables us to begin to make conclusions.

Let's consider another example. Jane's score in a test is 80. Assume that the mean score in this case is 100. We can see that Jane's score is 20 points below the mean. However, in saying that there is a difference of 20, we have not considered the deviation of **all scores from the mean**, so we still do not have a proper comparison. Knowing, for example, that Jane's score is one standard deviation below the mean still doesn't tell us enough. Unless we calculate all the

---

2 Derived from the article at the URL: http://wise.cgu.edu/sdtmod/reviewz.asp

z-scores, we cannot know for certain how Jane's score compares with the rest of the students in her class. Using the z-test, we can calculate a set of standard scores which are in terms of standard deviations from the mean of the dataset. By standardizing the scores, we can compare any individual score with all scores in a dataset.

SAMPLE SIZE FOR THE Z-SCORE:

In the Unit 3 you learned how to calculate the variance and the standard deviation. The larger the sample, the closer the variance (S_x^2) and standard deviation (S_x) are to that of the population. Generally you would use a sample size of at least 30. When the z-test is used on large samples and the distribution is normal, then the sample's standard deviation (S_x) is close to the population's standard deviation (**sigma or σ**).

Looking at exam scores, you can tell, for example, if Mary's math score is above or below the mean just by comparing her raw score with the calculated mean. By converting all the exam scores to z-scores, you can tell how many standard deviations Mary's z-score is away from the mean. Suppose her raw score is below the mean. If you know where Mary's score lies in reference to every other score, you can plot the scores onto a normal curve and see where Mary's score is relative to the rest of the scores in the class which can help determine her grade. Ideally, this is what the phrase "grading on the curve" should be mean.

Although the z-test should be used with large samples, the examples in this unit use small sets of observations. The process for calculating a z-score is the same regardless of the size of the dataset **but the validity of the z-score is much greater with large samples**. The smaller sample size does make the process more easily followed. However, in real life, to find a z-score, you need a large enough sample so that both the calculated mean and standard deviation are probably very close to the population's mean and standard deviation. Most undergraduate classes have at least 200 students.

The **z-scores** are found by subtracting the mean from **each** raw score of your observations then dividing the result by the standard deviation of the sample. So the first step in the z-test is to calculate the mean, variance and standard deviation.

The mean is the pivotal point of the data. Because the mean is being subtracted from the raw scores, the z-scores will be either **positive**, **negative**, or **zero**. A z-score less than zero shows an observation that is less than the mean and will be negative. A z-score greater than zero represents an observation that is greater than the mean and will be positive. If the z-score is zero, it represents an observation that is equal to the mean.

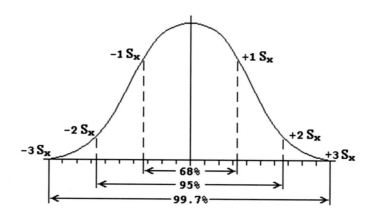

Figure 21: In a normal distribution, there are three standard deviations, S$_{x}$, to the right of the mean and three to the left.

By the ***empirical rule***, about 95% of the scores are within two standard deviations of the mean. So, just over 95% of the z-scores lie between -2 and +2 standard deviations and just over 99% are within 3 standard deviations. Any z-score outside that range is very unlikely (less than 1% chance of occurring).

UNIT 4B-1: BASIC PRINCIPLES FOR CALCULATING THE Z-SCORE

PURPOSE:

A **z-score** is also known as a **standard score.** Its main use is with large samples (30 or more) when a standardized score is needed; standardizing the score produces a normal distribution. If your dataset cannot be graphed as a normal curve, by calculating the z-scores, the resulting curve shows a normal distribution.

For a particular set of data, there are a variety of calculations to obtain a **z-score.** The z-score gives you a method for comparing one or two means. The purpose of this unit is to show you how to convert a raw score into a z-score with a given sample **mean** and a given **standard deviation,** then to determine if the result enables you to reject the null hypothesis, H_0. Two methods can be used:

1. Look up the table to find the **P-value** for the **calculated z-score**, for a predetermined significance level, usually 0.01 or 0.05. *If the P-value is less than the significance level, reject H_0.*

2. Determine the **critical value** of the z-score based on the level of significance. *If the calculated z-value is greater than the critical value, then reject H_0.*

OBJECTIVES:

- Convert data with a single mean to a set of z-scores.

- Compare two means by calculating the z-score.

- Standardize raw data for a distribution curve by converting the raw scores to z-scores.

- Calculate a test statistic using the raw score the mean and the standard deviation for a particular set of data.

- Compare the **critical value** with the **calculated value** and use these to determine if the null hypothesis can be rejected.

- Given a sample mean and the sample's standard deviation, calculate a set of z-scores and use a number line to illustrate it.

- Describe the use of a z-score with a single mean and with two means.

INTRODUCTION:

When we collect a set of observations, we can plot a distribution curve similar to the one in Figure 20. However, when collecting data during a research project, the distribution curve for sample data may or may not provide a normal curve.

Because the z-test standardizes the raw scores, it provides a reliable measure of the variable that is under consideration.

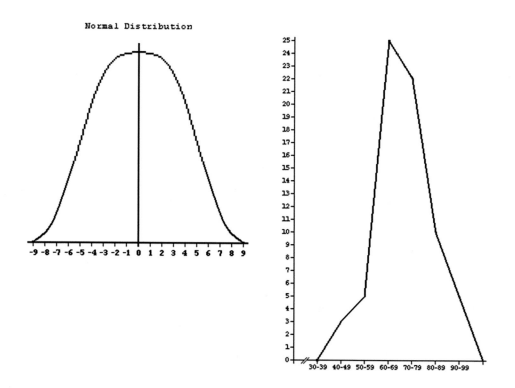

Figure 22: a normal distribution compared with a likely research data distribution.

The four-step process for testing hypotheses:

1. State the null and alternative hypotheses.

2. Formulate an analysis plan.

   a.   Analyze sample data.

   b.   Find the critical z-value.

3. Compare the calculated z-score with the critical z-value.

4. Interpret the results.

RAW SCORE TO Z-SCORE:

The z-scores are not affected by the distribution of the raw scores because they transform the raw score distribution into a normal distribution. The formulas for the z-test:

$z = \dfrac{X - \bar{X}}{S_x}$ The formula to convert **a raw score to a z-score:** where **X** is the raw score; $\bar{X}$ is the sample mean; and S_x is the standard deviation. The calculation is repeated for **all raw scores**.

$z = \dfrac{X - \bar{X}}{S_x / \sqrt{n}}$ **The test statistic:** where **X** is the raw score, $\bar{X}$ is the sample mean, S_x is the standard deviation, **n** is the number of observations and **z** is the standard score. The test statistic is the value calculated to determine the strength of the null hypothesis.

$S_x = \sqrt{\dfrac{\sum (X_i - \bar{X})}{n-1}}$ Where X_i represents each of **n** observations, 1,2,3,...,n; that is, the calculation is repeated for each of the **n** observations.

ALGORITHM FOR CALCULATING THE Z-SCORE TEST STATISTIC:

1. Determine the mean of the sample, $\bar{X} = \Sigma X / n$.

2. Subtract the mean from each of the raw scores to get the deviations, $(X - \bar{X})$.

3. Calculate the standard deviation as follows:

    a. Find the variance:

    b. Square each deviation $(X - \bar{X})^2$.

    c. Sum the squared deviations, $\Sigma (X - 0)^2$.

    d. Divide the sum of the squared deviations by **n-1, $\Sigma (X - \bar{X})^2 \div (n-1)$.**

    e. Find the square root of the variance to determine the standard deviation S_x of the sample. $\sqrt{\{\Sigma (X - \bar{X})^2 \div (n-1)\}}$

4. Divide the standard deviation by $\sqrt{n}$.

5. Divide each difference between mean and score $(X - \bar{X})$ by **the result of step 4**. The final result is the **z-score for each raw score**.

Symbols used in the algorithm:

S_x	The standard deviation's symbol for all observations **X.**
X	Each individual observation (raw score).
N	The number of observations in the set.
$\bar{X}$	The mean for the set of observations (the sample mean):
Σ	The symbol meaning "the sum of"
ΣX	Sum of all observations.

$X-\bar{X}$ Deviation of each individual score from the mean.

$(X-\bar{X})\div S_x$ Each deviation divided by the standard deviation for this set of observations.

$(X-\bar{X})^2$ Each deviation squared.

$\Sigma(X - \bar{X})^2$ The sum of all squared deviations.

EXAMPLE 1:

PROBLEM:

At the beginning of the school year, eight 4-year-old children were observed, to determine how many letters of the alphabet they could identify. Data for the previous five years shows that the intake of 4-year old pre-school children knew an average number of five letters.

ANALYSIS PLAN:

Convert the following observations into Z-scores. Determine if this year's intake is the same as the previous five years. The null hypothesis assumes this year's group is no different.

NULL & ALTERNATIVE HYPOTHESES:

$H_0: \bar{X} = \mu$ H_0 shows this is a two-tailed test.

$H_a: 0 \neq \mu$

LEVEL OF SIGNIFICANCE (A):

0.05

RAW SCORES (X):

8, 3, 2, 5, 7, 4, 10, 9 (don't sort the scores).

n-1 = 8=1 =7

Raw Scores X
8
3
2
5
7
4
10
9
n = 8
$(\Sigma X) = 48$

Figure 23a: Raw scores.

CALCULATION:

STEP 1:

Calculate the mean:

$$\overline{X}/n = 48/8$$

$$= 6$$

Raw Scores X	Deviations x-$\overline{X}$
8	8 - 6 = 2
3	3 - 6 = -3
2	2 - 6 = -4
5	5 - 6 = -1
7	7 - 6 = 1
4	4 - 6 = -2
10	10 - 6 = 4
9	9 - 6 = 3
n = 8	
(Σx) = 48	

Figure 23b: UNIT 4B-1, Example 1, Step 1, Calculate the mean.

STEP 2:

Calculate each deviation ($X-\overline{X}$) from the mean (column 2) and the variance ($X-\overline{X}$)^2.

Now you know that the sum of the deviations is equal to zero. To find the variance, you must square each deviation.

Raw Scores X	Deviations x-$\overline{X}$	Variance (x-$\overline{X}$)^2
8	8 - 6 = 2	4
3	3 - 6 = -3	9
2	2 - 6 = -4	16
5	5 - 6 = -1	1
7	7 - 6 = 1	1
4	4 - 6 = -2	4
10	10 - 6 = 4	16
9	9 - 6 = 3	9
n = 8		
(Σx)=48	$\overline{X}$ = 6	Σ(x-$\overline{X}$)^2 = 60

Figure 23c: UNIT 4B-1, example 1, step 2, calculate the deviations.

STEP 3:

Calculate the standard deviation. [Values rounded to two decimal places.]

$$S_x = \sqrt{[\Sigma(X-\overline{X})^2 \div n-1)]}$$

$$= \sqrt{60/7}$$

$$= \sqrt{8.57}$$

$$= 2.93$$

STEP 4:

Raw Scores X	Deviations x-$\overline{X}$	Variance (x-$\overline{X}$)^2	z-score (x-$\overline{X}$) $\div$ S_x
8	8 - 6 = 2	4	2$\div$2.93 = 0.68
3	3 - 6 = -3	9	-3$\div$2.93 = -1.02
2	2 - 6 = -4	16	-4$\div$2.93 = -1.37
5	5 - 6 = -1	1	-1$\div$2.93 = 0.34
7	7 - 6 = 1	1	1$\div$2.93 = -0.34
4	4 - 6 = -2	4	-2$\div$2.93 = -0.68
10	10 - 6 = 4	16	4$\div$2.93 = -1.37
9	9 - 6 = 3	9	3$\div$2.93 = -1.02
n = 8	(Σx)/n = $\overline{X}$ = 6	Σ(x-$\overline{X}$)^2 = 60	Σ(x-$\overline{X}$)^2 $\div$ (n-1)=S_x
(Σx) = 48			S_x = 2.93

Divide each deviation ($X-\overline{X}$) by the standard deviation for each raw score (S_x). This gives the position of each z-score in terms of standard deviations.

Figure 23d: UNIT 4B-1, e.g. 1, step 4, z-scores.

Graphing raw scores and the resulting z-scores onto a pair of number lines shows how the z-scores are evenly distributed around the mean. In fact, the z-score tells how far a particular observation is away from the mean. So, with a mean of 6, '10' is 1.41 standard deviations to the **right** of the mean. And '2' is -1.41 standard deviations to the **left** of the mean.

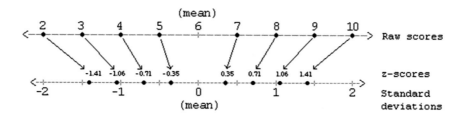

Figure 24: raw scores compared to z-scores.

The z-score for the mean is always zero. The second line shows the z-scores in comparison to the mean of zero (0), in terms of standard deviations. Notice that there is a slight shift inward from the mean in the placement of the z-scores with reference to their originating raw scores. This is **not** significant as the shift is determined by the scale used for the standard deviation's number line. What **is** significant are the signs +ve (abbreviation for 'positive') and -ve (abbreviation for 'negative') which **are** highly significant. The sign shows the position (right or left of the mean) and the number shows the magnitude (size) of the distance from the mean. Converting raw scores to z-scores (standard scores) can be regarded as a "normalizing" process, that is, the curve representing the z-scores is a normal or bell-shaped curve even though the raw scores may or may not have a normal distribution.

We have established that the mean of our sample is 6, whereas the hypothesized population mean was 5. [Note: had the population mean been a different number, the results would be different.] Does this difference mean that the sample's mean is significantly different from the hypothesized mean? **NO.** We need to determine the P-value or the critical z-score.

THE P-VALUE:

The P-value tells us how much evidence there is to allow us to reject the null hypothesis. Since H_0: $\overline{X}$ = μ, this is a two-tail test expressed by the limiting values of the z-scores, -1.41 and +1.41. This is usually written as ±1.41, that is, "plus or minus 1.41".
The **P-value** of the z-score gives us the area (a) under the curve **between the mean and the value of the z-score.** This shows the null hypothesis's limiting values. So the area of the white region in Figure 25 is 2a. The rejection regions, blue, in a two-tailed test are each equal to half of a.

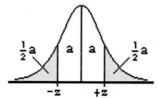

Figure 25: The rejection regions, blue, are each equal to half of 'a'

FACTS ABOUT THE P-VALUE:

1. The probability of any event is a number between 0 and 1.

2. The sum of the probabilities or outcomes in a sample space is 1.

3. The total area under the curve is 1.

4. The closer to 1 the probability is, the more likely is the occurrence of the event.

5. P(A) + P(not A) = 1, that is, P(not A) is the complement of P(A) just as H_a is the complement of H_0.

Z	0.00	0.01	0.02	0.03	0.04	0.05	0.06	0.07	0.08	0.09
0.0	0.0000	0.0040	0.0080	0.0120	0.0160	0.0199	0.0239	0.0279	0.0319	0.0359
0.1	0.0398	0.0438	0.0478	0.0517	0.0557	0.0596	0.0636	0.0675	0.0714	0.0753
0.2	0.0793	0.0832	0.0871	0.0910	0.0948	0.0987	0.1026	0.1064	0.1103	0.1141
0.3	0.1179	0.1217	0.1255	0.1293	0.1331	0.1368	0.1406	0.1443	0.1480	0.1517
0.4	0.1554	0.1591	0.1628	0.1664	0.1700	0.1736	0.1772	0.1808	0.1844	0.1879
0.5	0.1915	0.1950	0.1985	0.2019	0.2054	0.2088	0.2123	0.2157	0.2190	0.2224
0.6	0.2257	0.2291	0.2324	0.2357	0.2389	0.2422	0.2454	0.2486	0.2517	0.2549
0.7	0.2580	0.2611	0.2642	0.2673	0.2704	0.2734	0.2764	0.2794	0.2823	0.2852
0.8	0.2881	0.2910	0.2939	0.2967	0.2995	0.3023	0.3051	0.3078	0.3106	0.3133
0.9	0.3159	0.3186	0.3212	0.3238	0.3264	0.3289	0.3315	0.3340	0.3365	0.3389
1.0	0.3413	0.3438	0.3461	0.3485	0.3508	0.3531	0.3554	0.3577	0.3599	0.3621
1.1	0.3643	0.3665	0.3686	0.3708	0.3729	0.3749	0.3770	0.3790	0.3810	0.3830
1.2	0.3849	0.3869	0.3888	0.3907	0.3925	0.3944	0.3962	0.3980	0.3997	0.4015
1.3	0.4032	0.4049	0.4066	0.4082	0.4099	0.4115	0.4131	0.4147	0.4162	0.4177
1.4	0.4192	0.4207	0.4222	0.4236	0.4251	0.4265	0.4279	0.4292	0.4306	0.4319
1.5	0.4332	0.4345	0.4357	0.4370	0.4382	0.4394	0.4406	0.4418	0.4429	0.4441

Figure 26: The z-table is used to find the area of a z-score of 1.4.

To reject the null hypothesis the area under the curve to the left of -z or the area under the curve to the right of +z must be less than the level of significance.

$H_0: \overline{X} = \mu$

$H_a: \overline{X} \neq \mu$

The area under the curve for a particular z-score is found by examining the z-table. The area for a z-score of 1.37 (rounded to 1.4) is 0.4207 (circled). This means that the area under the curve between 1.41 (or -1.41) and the mean is 0.4207.

The total area is:

1 - (2 x 0.4207)

= 1 - 0.8414

= 0.1586.

The significance level selected was 0.05 which is less than 0.1586. That means we cannot reject the null hypothesis. This suggests that this year's intake of 4-year old children in spite of the different number of letters known is not significantly different from the intake of previous years.

The following URL takes you to a site where you can use the above table to determine the percentage of the population under any curve.

http://www.mathsisfun.com/data/standard-normal-distribution-table.html

WHAT USE IS THE AREA UNDER THE CURVE?

Example 1 shows how to convert the raw scores to z-scores. If you have a z-score, you can calculate the percentage of respondents who are in a particular area of the curve (it works best for large samples). This table[3] can be used to find the area under the curve (see Figure 4) from the vertical axis at zero (the mean) to the value of the z-score.

The question is: Why would we want to know the area under the curve? It tells us what segment of the population is within the area, between the mean and 1.41 or -1.41. That area is **a**, with **a** = 0.4207. That means that about 84% of the study population lies in the un-shaded area of the curve and about 16 percent lie in the two shaded area (approximately 8% in each shaded area).

You can also determine where a particular individual lies on the standard normal curve in order to assign a grade. This also works best with large samples such as you have in university undergraduate courses. However, we have not done anything to determine the **significance** of the information we have calculated. To do that we must consider the **test statistic and the critical z value; the symbol for the critical z value is** z*.

3 Pierce, Rod. "Standard Normal Distribution Table" Maths Is Fun. Ed. Rod Pierce. 30 Jun 2009. Used with permission.

CRITICAL VALUES:

The way to determine if the null hypothesis, H_0 can be rejected is through using the critical z value. First you must calculate the **z-value** which is not the same as the z-scores calculated in the example. Then the critical z-value is taken from a table which gives the critical values through the chosen confidence levels. ***There is only one critical z-value for each level of confidence. If you are going to do a lot with z-scores, the table below can be memorized.***

CALCULATED z-VALUE:

S_x = **2.93**

$\sqrt{n}$ = $\sqrt{8}$ = 2.83

$S_x \div \sqrt{n}$ = 0.33   to 2 decimal places.

If μ = 5 (given), and **0** = 6 (calculated)

z = $\dfrac{6-5}{0.33}$

= $\dfrac{1}{0.33}$

= 3.03 (calculated z)

The critical z-value (z* as shown in the Confidence Table) for the 95% (α = 0.5) is 1.96; so the calculated value > critical value.

Rule: If the z-score is greater than the critical value, the null hypothesis can be rejected. If it is less, then we must fail to reject the null hypothesis.

Since the calculated value of 3.03 is greater than the critical value of 1.96 at 95%, the mean of the sample is significantly different from the hypothesized mean. This year's intake is different from previous intakes.

CONFIDENCE LEVEL	z* value
90%	1.645
95%	1.96
98%	2.33
99%	2.58

Figure 27: There is a single critical value, z*, for each confidence level, 90% through 99%.

There is only one critical z-value for each level of confidence; that is the critical values are fixed. If you are going to do a lot with z-scores, the table can be memorized.

[This page left blank to maintain correct pagination.]

UNIT 4B-2: USING THE Z-TEST TO ANALYZE MEANS

PURPOSE:

When using the z-test to analyze means, you have one of two situations. You have a single mean 9one variable) and must compare that mean with the population mean or you have two means and need to compare the two variables.

The z-score can be likened to a yardstick for a normal distribution. In describing the position of a particular raw score, the z-score is sometimes called a ***normal deviate*** as it describes how much the position of the raw score deviates from the mean. The sample size for the normal distribution ideally should be at least 30 with a single variable.

Figure 28: There are two paths when using means to analyze z-scores.

OBJECTIVES:

- Calculate a z-score when your data has only a single mean.

- Calculate a z-score when your data has two means.

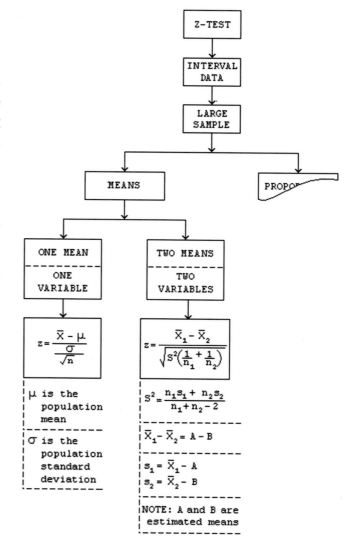

Z-TEST

INTERVAL DATA

LARGE SAMPLE

MEANS — PROPO

ONE MEAN / ONE VARIABLE

$$z = \dfrac{\bar{X} - \mu}{\dfrac{\sigma}{\sqrt{n}}}$$

μ is the population mean

σ is the population standard deviation

TWO MEANS / TWO VARIABLES

$$z = \dfrac{\bar{X}_1 - \bar{X}_2}{\sqrt{S^2\left(\dfrac{1}{n_1} + \dfrac{1}{n_2}\right)}}$$

$$S^2 = \dfrac{n_1 s_1 + n_2 s_2}{n_1 + n_2 - 2}$$

$\bar{X}_1 - \bar{X}_2 = A - B$

$s_1 = \bar{X}_1 - A$
$s_2 = \bar{X}_2 - B$

NOTE: A and B are estimated means

A. TESTING A SINGLE MEAN:

We'll first look at the z-test when it is used to analyze a single mean, that is, you only have one data set with the one variable, **X**. Your intent is to compare the sample mean for the data set with the population mean or with a hypothesized mean.

The test statistic formula:

$$z = \frac{\overline{X} - \mu}{\frac{\sigma}{\sqrt{n}}}$$

In this formula, **σ/√n** is the **standard error**. The standard error is an estimate of the population standard deviation. In most cases the population standard deviation is not available; instead, we can use a sample standard deviation, **S**.

If **σ** is the population standard deviation then **S$\overline{X}$** is the sample estimate of **σ**. When the mean of the sample is **$\overline{X}$**, then the standard error for that mean is, by definition:

$$SE_{\overline{x}} = \frac{S_{\overline{x}}}{\sqrt{n}}$$

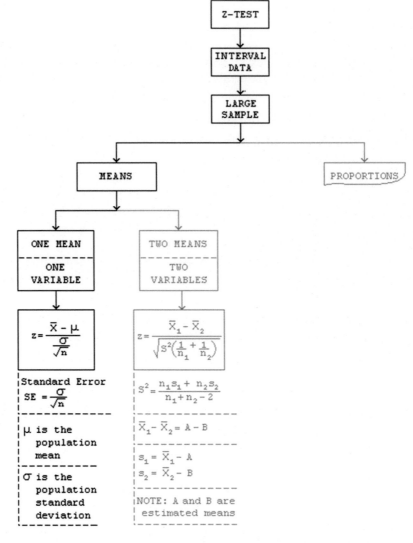

Figure 29: Emphasizing the path for analyzing a single variable.

Note that the standard* error *is a measure of variability not a measure of central tendency.

As a general rule, we do not use a frequency table when determining a z-score. With thirty or more and maybe as many as 200 or 300 observations in the sample, doing individual scores could become cumbersome. Fortunately, there is an alternative: The z-scores reference the individual **exam scores.** The professor wants a mean for the range of the exam scores **not** the average score for all students in his class; in other words, ***the frequencies are irrelevant for calculating the z-scores***.

UNIT 4B-2 EXAMPLE 1. THE PROCESS FOR CALCULATING THE Z-SCORES:

Given the following data set, find the frequency, the mean and the standard deviation

SCORE	FREQUENCY
30	5
31	7
32	8
33	10
34	12
35	15
36	19
37	20
38	18
39	17
40	19
41	21
42	25
43	13
44	10
45	16
n=16	

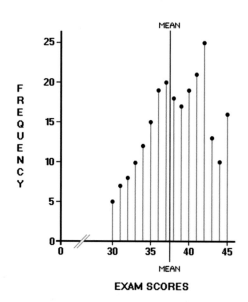

Figure 30: Data is in the form of a frequency table and frequency graph.

STEP 1:

To calculate the mean, **do not use** frequencies. You must use the sum of the scores and divide by **n**. In this case, **n** is equal to 16. The sum of the scores is 600, so the mean is:

$\overline{X}$ = 600÷16
= 37.5

STEP 2:

Calculate **X-00** for **each** exam score column 3).

STEP 3:

Calculate $(X-\overline{X})^2$ for each score (column 4):

STEP 4:

Calculate **the sum of all** $(X-\overline{X})^2$.

$\sum(X-\overline{X})^2$ = 340

STEP 5:

Score (X)	FR	$x - \overline{X}$	$(x - \overline{X})^2$
30	5	30 – 37.5 = –7.5	56.25
31	7	31 – 37.5 = –6.5	42.25
32	8	32 – 37.5 = –5.5	30.25
33	10	33 – 37.5 = –4.5	20.25
34	12	34 – 37.5 = –3.5	12.25
35	15	35 – 37.5 = –2.5	6.25
36	19	36 – 37.5 = –1.5	2.25
37	20	37 – 37.5 = –0.5	0.25
38	18	38 – 37.5 = +0.5	0.25
39	17	39 – 37.5 = +1.5	2.25
40	19	40 – 37.5 = +2.5	6.25
41	21	41 – 37.5 = +3.5	12.25
42	25	42 – 37.5 = +4.5	20.25
43	13	43 – 37.5 = +5.5	30.25
44	10	44 – 37.5 = +6.5	42.25
45	16	45 – 37.5 = +7.5	56.25
n = 16		$\overline{X}$ = 37.5	$\sum(x-\overline{X})^2$ = 340

Figure 31: Frequency column is shaded yellow to remind you the frequencies are NOT used in the calculation.

The **variance**, S_x, is also based on the exam scores and the mean not the frequencies.

$S_x^2 = \{\Sigma(X - \overline{X})^2\}\div(n-1)$

$= 340/15$

$= 22.67$

$S_x = \sqrt{(S_x^2)}$

$= \sqrt{(22.67C}$

$= 4.76$

Remember the z-score of any mean is always zero.

STEP 6:

Z-scores are found by dividing $(X-\overline{X})/S_x$ as shown in column 4.

SC (X)	x - X̄	(x - X̄) ÷ Sₓ	z-scores
30	30 − 37.5 = −7.5	−7.5 ÷ 4.76 =	−1.58
31	31 − 37.5 = −6.5	−6.5 ÷ 4.76 =	−1.37
32	32 − 37.5 = −5.5	−5.5 ÷ 4.76 =	−1.16
33	33 − 37.5 = −4.5	−4.5 ÷ 4.76 =	−0.95
34	34 − 37.5 = −3.5	−3.5 ÷ 4.76 =	−0.74
35	35 − 37.5 = −2.5	−2.5 ÷ 4.76 =	−0.53
36	36 − 37.5 = −1.5	−1.5 ÷ 4.76 =	−0.32
37	37 − 37.5 = −0.5	−0.5 ÷ 4.76 =	−0.11
38	38 − 37.5 = +0.5	+0.5 ÷ 4.76 =	+0.11
39	39 − 37.5 = +1.5	+1.5 ÷ 4.76 =	+0.32
40	40 − 37.5 = +2.5	+2.5 ÷ 4.76 =	+0.53
41	41 − 37.5 = +3.5	+3.5 ÷ 4.76 =	+0.74
42	42 − 37.5 = +4.5	+4.5 ÷ 4.76 =	+0.95
43	43 − 37.5 = +5.5	+5.5 ÷ 4.76 =	+1.16
44	44 − 37.5 = +6.5	+6.5 ÷ 4.76 =	+1.37
45	45 − 37.5 = +7.5	+7.5 ÷ 4.76 =	+1.58
n = 16	X̄ = 37.5	Sₓ = 4.76	

Figure 32: Step 6 - calculate the z-scores

ONE USE FOR Z-SCORES: GRADING ON THE CURVE:

There is one use of z-scores that is not be obvious: using them to calculate the grades of an undergraduate class. This is particularly useful as most general education undergraduate classes have 200 or more participants. We will use the z-scores from Figure 32. **Note:** A set of z-scores is a set of standardized scores which indicates that a graph of these scores will show as a bell curve. As a result the bell curve graph can be used to determine final scores

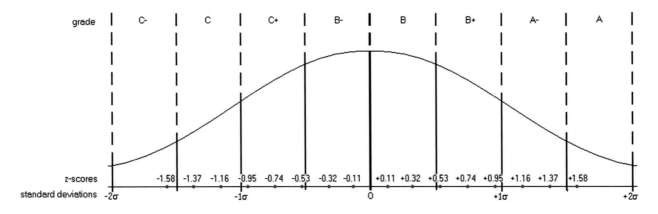

Figure 33: Using the position of the standard deviations as "separators" for the grades, you can calculate the grades on the curve.

NOTE: Although the original scores are not in a normal distribution, because the z-scores are standardized, the distribution of the z-scores has a normal distribution.

The curve in Figure 33 shows a normal distribution between -2σ and +2σ. Placement of the z-scores shows the possible grade that can be assigned to each score. The grades are "possible" because each professor chooses his own grading method. For instance, the final grade in one class was actually based on the mid-term exam results. One student received a B+ on the mid-term and an A on the final. His final grade was B+. He protested that his final exam score was good enough to have averaged with the mid-term to an A-, but the professor's excuse for not changing his grade was because he would have to do it for others as well! The student asked, "Suppose one student got only a C on the midterm but improved enough to get an A on the final, would you give that student just a C?" There was no reply. Hopefully that professor will use a different method to grade his students in the future.

UNIT 4B-2 EXAMPLE 2, PROCESS FOR ANALYZING A SINGLE MEAN.

The data in Figure 32 will be used to determine if there is a significant difference between the mean and the historical or population mean. Assume that, historically, the **mean** for this particular exam has been **40** points. This value will be used as **μ, the population mean**.

STEP 1. STATE THE HYPOTHESES:

$H_0: \overline{X} = \mu$ The null hypothesis states the mean for the current test is the same as the historical value (fail to reject **H₀**).

$H_a: \overline{X} \neq \mu$ The alternative hypothesis states the mean for this test is not equal to the historical mean (reject H₀).

STEP 2. FORMULATE AN ANALYSIS PLAN:

Use the z-test for a single mean.

$$z = \frac{\overline{X} - \mu}{\frac{\sigma}{\sqrt{n}}}$$

STEP 3. ANALYZE SAMPLE DATA:

The z-score:

$$z = \frac{\overline{X} - \mu}{S_{\overline{x}} \div \sqrt{n}}$$ **wh**ere $\overline{X}$ = 37.5, **μ** = 40, $S_{\overline{x}}$ = 47.6, **n** = 16, $\sqrt{n}$ = 4

$$z = \frac{37.5 - 40}{47.6 \div 4}$$

$$= \frac{-2.5}{11.9}$$

$$= -0.21$$

Significance level at 0.05 (95%) = 1.96

Calculated z < critical z

-0.21 < 1.96 **0** Fail to reject **H₀**

STEP 4. INTERPRET THE RESULTS:

Conclusion: the scores of this year's undergraduates are not significantly different from the historical scores.

CONFIDENCE LEVEL	z* value
90%	1.645
95%	1.96
98%	2.33
99%	2.58

Figure 34: z* at 95% confidence level.

TESTING TWO MEANS:

Two means must be treated differently. The z-score for **each** dataset's observations can be calculated as individual scores within each dataset using the means ($\overline{\mathbf{X}}_1$, $\overline{\mathbf{X}}_2$), but when you wish to compare the two sets, the formula is:

$$z = (\overline{\mathbf{X}}_1 - \overline{\mathbf{X}}_{2)} \div \sqrt{[\sigma^2(1/n_1 + 1/n_2)]}$$

$$\text{where } \sigma^2 = (n_1 S_1^2 + n_2 S_2^2) \div (n_1 + n_{2)}$$

When the population standard deviation is not available, we can use the sample standard deviation, $\mathbf{S_x}$. The variance for each dataset is calculated as follows:

$$\mathbf{S_1} = \mathbf{X_1 - A}$$

$$\mathbf{S_2} = \mathbf{X_2 - B}$$ Where A and B are the estimated means for each dataset.

HOW DO YOU FIGURE OUT THE ESTIMATED MEANS?

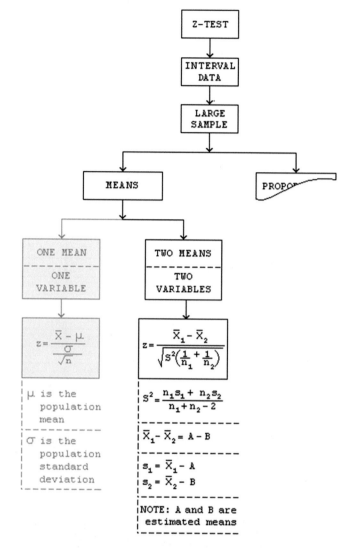

Figure 35: The path for analyzing two means is emphasized.

UNIT 4B-2 EXAMPLE 3. USING ESTIMATED VALUES FOR THE MEANS:

NOTE: While it isn't essential to order your data (not done in this example), you may prefer to do so.

In this problem, the nine participants were tested (X_1) before a brief intensive training to improve their understanding of specific concepts and were tested again afterwards (X_2). The null hypothesis states that before and after scores would be the same (that is, no significant difference).

It was decided to use estimated values for the means. Why? Actual means usually have decimal values. An estimated mean allows you to use a whole number, simplifying the calculation, particularly when working with larger quantities.

Students	Before (X_1)	$X_1 - \bar{X}_1$	$(X_1 - \bar{X}_1)^2$	After (X_2)	$X_2 - \bar{X}_2$	$(X_1 - \bar{X}_1)^2$
1	1900	0	0	2100	0	0
2	1800	−100	10000	1900	−200	40000
3	2000	100	10000	2250	150	22500
4	1800	−100	10000	1950	−50	2500
5	1900 A	0	0	1950	−150	22500
6	2000	100	10000	2100 B	0	0
7	1700	200	40000	1950	−150	22500
8	2000	100	10000	2350	250	62500
9	1900	0	0	2150	50	2500
n = 9	Σ = 17000		$\Sigma(X_1 - \bar{X}_1)^2$ = 90000	Σ = 16750		$\Sigma(X_1 - \bar{X}_1)^2$ = 175000

Figure 36: Selecting estimated means.

For an estimated value of the mean, select a value for **X** that is close to the center of your dataset. In Figure 36, the selected estimated means (highlighted) are shown by **A** for the first dataset and **B** for the second. The decision is fairly arbitrary.

Calculate the standard deviations for each dataset as shown. Since the information is based on estimated means, we also have **_estimated standard deviations_**; 'n' for both datasets is 9.

Estimated Mean (**A**) = 1900

$$S_1^2 = \{\Sigma(X_1 - \bar{X}_1)^2 \div (n-1)\}$$
$$= 90000/8$$
$$= 11250$$

Estimated Mean (**B**) = 2100

$$S_2^2 = \{\Sigma(X_2 - \bar{X}_2)^2 \div (n-1)\}$$
$$= 172500/8$$
$$= 215625.5$$

$$\sigma = \frac{n_1 S_1^2 + n_2 S_2^2}{n_1 + n_2} \qquad = \frac{1900 - 2100}{\sqrt{16562.5\left(\frac{1}{9} - \frac{1}{9}\right)}}$$

$$= \frac{200}{\sqrt{16562.5\left(\frac{2}{9}\right)}}$$

$$= \frac{200}{60.67}$$

$$= 3.30$$

The critical value of **z** (**z***) at 99%

z* = 2.58

Calculated z > critical z

3.30 > 2.58

Reject the null hypothesis.

CONFIDENCE LEVEL	z* value
90%	1.645
95%	1.96
98%	2.33
99%	2.58

Figure 37: Critical z at 99% level.

Conclusion:

The students' scores after receiving the intensive training are significantly different from their scores before receiving the training. The training was effective.

UNIT 4B-3: USING THE Z-TEST TO ANALYZE PROPORTIONS

PURPOSE:

A proportion is a part of a whole considered in relation to the whole of which it is a part. The proportion can be written as **ratio** (a:b), as a **fraction** (a/b), or as a percentage (a/b*100), where **b** is the whole and **a** is the part. The "whole" is a complete dataset with a common characteristic or attribute.

An example would be, suppose we are considering the proportion of pets owned by a friend which have four legs. He has a cat, a dog, a bird and fish. Only the cat and dog have four legs, so the proportion can be written as: 2 out of four or 2:4; 2/4 or 0.5; or 50%

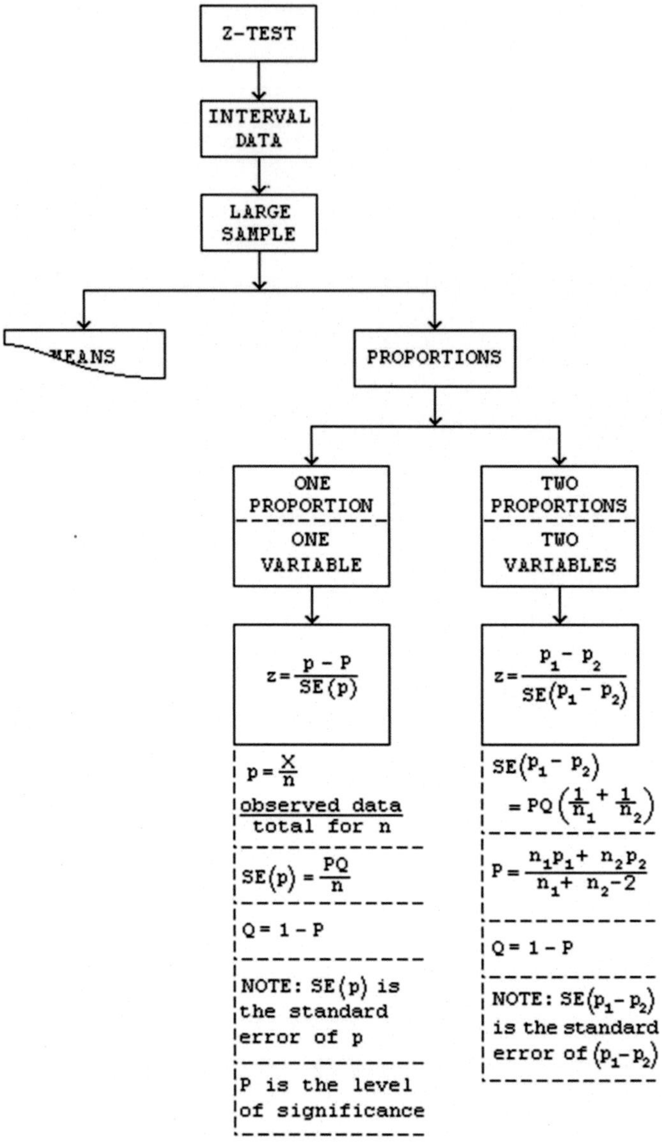

Figure 38: Paths for analyzing proportions.

Flowchart:

Z-TEST → INTERVAL DATA → LARGE SAMPLE → branches to MEANS and PROPORTIONS

PROPORTIONS branches to:

ONE PROPORTION / ONE VARIABLE

$$z = \frac{p - P}{SE(p)}$$

$p = \frac{X}{n}$ — $\underline{\text{observed data}}$ / total for n

$SE(p) = \frac{PQ}{n}$

$Q = 1 - P$

NOTE: $SE(p)$ is the standard error of p

P is the level of significance

TWO PROPORTIONS / TWO VARIABLES

$$z = \frac{P_1 - P_2}{SE(P_1 - P_2)}$$

$SE(P_1 - P_2) = PQ\left(\frac{1}{n_1} + \frac{1}{n_2}\right)$

$P = \frac{n_1 P_1 + n_2 P_2}{n_1 + n_2 - 2}$

$Q = 1 - P$

NOTE: $SE(P_1 - P_2)$ is the standard error of $(P_1 - P_2)$

OBJECTIVES:

- Given a single variable, determine the P-value for the rejection region in a one-tailed test and decide if H_0 can be rejected.

- Given two variables, determine the P-value for the rejection region in a two-tailed test and decide if H_0 can be rejected.

- Given your own interval data, calculate the set of z-scores and illustrate the result and determine if H_0 can be rejected.

SUCCESS AND FAILURE:

When we consider probability, we are actually looking at the probability of success or failure. Our sample would require us to have **at least ten successes and at least ten failures** for the z-test we use to be valid. How do we define success or failure?

Suppose a professor has created a new test and wishes to find out the probability of passing the test. He wants the test to be difficult but not impossible. Suppose he administers the test to one thousand students. Possibilities:

1. If 998 students failed and 2 passed (998:2), then there are less than ten successes, the test would not be valid. The test is too hard.

2. Suppose there were 4 failures and 996 passes (4:996). Because there are less than ten failures, the test is not valid. It is too easy.

3. For the test to be valid, there must be at least 10 successes that is, those who passed (10:990) AND at least 10 failures (10:990)

How can we tell if there are enough success and failures? The proportion itself will tell you. The forms of a proportion are **a ratio**, a fraction, or a percentage. Convert the proportion to a percentage and if it lies between 10% (10:100 or 1:10) and 90% (90:100 or 9:10) there are enough successes and failures to fulfill the requirement. [Remember that 1:2 is 50% and 1:4 is 25%.]

A. SINGLE PROPORTION:

The proportion refers to the fraction of the whole which has a common characteristic. For example, in a small city, 600 men were selected at random as a sample. On examination, 325 men were found to be non-smokers. Can you conclude that there are more non-smokers than smokers in the population?

The proportion (**P**) in this example is 325:600, that is, the sample compared with the sample total. The level of significance is **α** = 0.05. If we assume that the sample is truly representative of the population of men in the city, then we can make the comparison that if the proportion is significant using a z-value, then the majority of men in the city are non-smokers.

Highlighting the portion of the flow chart dealing with one proportion or one variable, note that SE(p) is the standard error of p and that P is the level of significance.

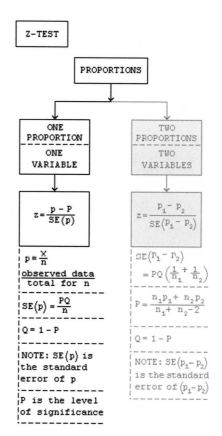

Figure 39: Emphasizing the path for a single proportion.

UNIT 4B-3. EXAMPLE 1:

In the City, there are 600 participants in the study. Of these, 325 are non-smokers. Is the proportion of smokers to non-smokers significant?

GIVEN DATA:

n = 600

X = 325

α = 0.05 (significance level)

(p = $X \div 600$ = $325 \div 600$ = 0.542 (where p is the proportion involved).

HYPOTHESES:

H_0: P = 0.5   [The number of smokers and non-smokers are equal.]

H_a: P > 0.5   [There are more non-smokers than smokers.]

SOLUTION:

The standard error of **SE(p)** is involved in this problem. This is also to be regarded as the probability of failure.

SE(p) = $\sqrt{(PQ/n)}$; where P = 0.5; Q = 1-P = 1-0.5

$$z = \frac{p - P}{SE(p)}$$

$$= \frac{0.542 - 0.05}{\sqrt{(0.5*0.5) \div 600}}$$

$$= \frac{0.492}{\sqrt{(0.25/600)}}$$

$$= \frac{0.492}{\sqrt{(0.00042)}}$$

$$= 0.0492 \div 0.0204$$

$$= 2.412$$

CONFIDENCE LEVEL	z* value
90%	1.645
95%	1.96
98%	2.33
99%	2.58

Figure 40: critical z for a confidence level of 95%.

*z = 1.96

Cal |z| of 2.412 > ***z**

CONCLUSION:

REJECT **H_0** therefore the conclusion is that the number of non-smokers in the city is significantly greater than the number of smokers. **NOTE:** The ratio of non-smokers to total participants is 325:600 which leaves 275 smokers. It seems obvious that there are more non-smokers than smokers. ***BUT just looking at the data does not determine if the difference is significant! That's why you must do the analysis.***

B. TESTING TWO PROPORTIONS:

When you have two proportions, the population size for each proportion may or may not be the same. Standard error is used in the calculation.

The null hypothesis, $H_0: P_1 = P_2$, says the two proportions are equal.

The alternative hypothesis $H_a: P_1 \neq P_2$, says the two proportions are not equal.

This becomes a two-tail test with the left tail being $P_1 < P_2$ and the right tail being $P_1 > P_2$. The level of significance may be $\alpha = 0.05$ or 0.01.

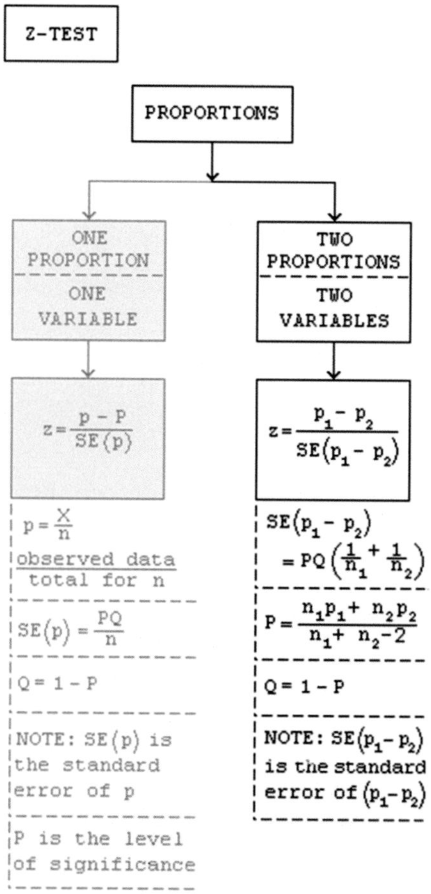

Figure 41: Unit 4B-2, emphasis on the path for two proportions.

The test statistic for two proportions is:

$$z = \frac{P_1 - P_2}{SE(P_1 - P_2)}$$

Where P_1 & P_2 are sample proportions; **SE** is the standard error and P is the probability of success

$$SE(p_1 - p_2) = \sqrt{PQ\left(\frac{1}{n_1} + \frac{1}{n_2}\right)}$$

$$P = \frac{n_1 P_1 + n_2 P_2}{n_1 + n_2}$$

$$Q = 1 - P$$

UNIT 4B-3. EXAMPLE 2:

CRITICAL VALUE AND ANALYSIS:

If Cal |z| > critical value, reject null hypothesis; if less than the critical value, fail to reject the null hypothesis. Reminder: the symbol of "Cal |z|" means that the **absolute value of z** is to be considered (that is, ignore any positive or negative sign).

THE PROBLEM:

The Klinger University hospital has tested a specific drug and determined that it is effective. Now they are conducting a test on the drug to determine if the drug has equal effects on males and females. The sample size for men is 200 and for women 150. P1 = 0.51 and P2 = 0.46 (for men); α = 0.05

HYPOTHESES:

$H_0: P_1 = P_2$

$H_a: P_1 \neq P_2$

SOLUTION:

$$z = \frac{P_1 - P_2}{SE(P_1 - P_2)}$$

$$SE(p_1 - p_2) = \sqrt{PQ\left(\frac{1}{n_1} + \frac{1}{n_2}\right)}$$

$$P = \frac{n_1 P_1 + n_2 P_2}{n_1 + n_2}$$

$$Q = 1 - P$$

CALCULATION:

$n_1 = 150$ **number of women in the sample.**

$n_2 = 200$ **number of men in the sample.**

$$P = \frac{150(0.51) + 200(0.46)}{150 + 200}$$

$$= \frac{76.5 + 92}{350}$$

$$= \frac{168.5}{350}$$

$$= 0.48$$

$$Q = 1 - P$$
$$= 1 - 0.48$$
$$= 0.52$$

$$SE(p_1 - p_2) = \sqrt{PQ\left(\frac{1}{n_1} + \frac{1}{n_2}\right)}$$

$$= \sqrt{PQ\left(\frac{1}{150} + \frac{1}{200}\right)}$$
$$= \sqrt{(0.48*0.52)(0.0067 + 0.005)} \quad \text{where } * = \text{multiply by}$$
$$= \sqrt{(0.2496)(0.0117)}$$
$$= \sqrt{(0.002920)}$$
$$= 0.05403$$

$$z = \frac{\bar{X}_1 - \bar{X}_2}{\sqrt{S^2\left(\frac{1}{n_1} + \frac{1}{n_2}\right)}}$$

$$z = \frac{0.51 - 0.46}{0.054}$$
$$= 0.006/0.054$$
$$= 0.9259 \text{ (calculated |z|)}$$

(Calculated |z|) = 0.9259

Critical z = $Z_{0.05}$ → *Z = 1.96

Calculated |z| < critical z

0.9259 < 1.96 → Fail to reject H_0

→ Drug is equally effective for men and women.

UNIT 4B. ASSIGNMENT: USING THE Z-TEST & Z-SCORE.

Respond to the following three questions and check the results.

1. Describe how the mean, standard deviation and the set of observations enable you to calculate the set of z-scores.

2. Calculate the z-scores for the following dataset.

   a.   14, 8, 9, 8, 12, 10, 7, 15, 11, 9

   b.   The population mean, µ, is 10, and the population standard deviation, σ, is 12.

3. If your data gives you a single mean, use the z-test to convert the raw scores to z-scores, and illustrate the result.

The four-step process for testing hypotheses:

1. State the null and alternative hypotheses.

2. Formulate an analysis plan.

   a.   Analyze sample data.
   b.   Find the critical z-value (*z).

3. Compare the calculated z-score (test statistic) with the critical z-value.

4. Interpret the results.

Algorithm for calculating z-scores.:

1. Determine the mean of the sample, $\overline{X} = \Sigma X/n$.

2. Subtract the mean from each of the raw scores to get the deviations, $(X-\overline{X})$.

3. Find the Variance as follows:

   ◦   Square each deviation $(X-\overline{X})^2$.
   ◦   Sum the squared deviations, $\Sigma(X-\overline{X})^2$.
   ◦   Divide the sum of the squared deviations by $n-1$, $\Sigma(X-\overline{X})^2 \div (n-1)$.

4. Find the square root of the variance to determine the standard deviation, Sx. $\sqrt{\{\Sigma(X-0)^2 \div (n-1)\}}$

5. Divide the standard deviation, Sx, by $\sqrt{n}$.

6. Divide each difference between mean and score $(X-\overline{X})$ by **the result of step 5**. The final result is the **z-score for each raw score.**

UNIT 4B. ASSIGNMENT FEEDBACK: USING THE Z-TEST & Z-SCORE.

Question 2. Calculate the z-scores for the following dataset.

7, 8, 9, 9, 6, 10, 7, 5, 9, 6

μ = 10 (given)

σ = 12 (given)

n = 10

STEP 1: MEAN.

$$\bar{X} = \sum X/n$$
$$= 76 \div 10$$
$$= 7.6$$

STEP 2: VARIANCE.

$$S_x^2 = \sum(X-\bar{X})^2 \div (n-1)$$
$$= 24.6 \div 9$$
$$= 2.73$$

SCORE	$(X-\bar{X})$	$(X-\bar{X}) \div \sqrt{n}$
7	7 - 7.6 = -0.6	-0.6 ÷ 3.16
8	8 - 7.6 = 0.4	0.4
9	9 - 7.6 = 1.4	1.4
9	9 - 7.6 = 1.4	1.4
6	6 - 7.6 = -1.6	-1.6
10	10 - 7.6 = 2.4	2.4
7	7 - 7.6 = -0.6	-0.6
5	5 - 7.6 = -2.6	-2.6
9	9 - 7.6 = 1.4	1.4
6	6 - 7.6 = -1.6	-1.6
$\sum\bar{X} = 76$ $n = 10$		$\sum(x-\bar{x})^2 = 24.6$

Figure 42a: Assignment data set.

STEP 3: STANDARD DEVIATION.

$$S_x = \sqrt{[\sum(X-\bar{X})^2 \div (n-1)]}$$
$$= \sqrt{S_x^2}$$
$$= \sqrt{2.73}$$
$$= 1.65 \text{ to 2 decimal places}$$

STEP 4: CALCULATE z.

Divide the standard deviation, Sx, by $\sqrt{n}$.

$$z = 1.65 \div \sqrt{10}$$
$$= 1.65 \div 3.16 \text{ (to 2 decimal places)}$$
$$= 0.52 \qquad \text{(to 2 decimal places)}$$

STEP 5: FIND THE z-SCORES

Divide each difference between mean and score **(X-$\overline{X}$)** by **the result of step 4**. The final result is the **z-score for each raw score.**

SCORE	(X-$\overline{X}$)	(X-$\overline{X}$)÷(S_x÷$\sqrt{n}$)	Z-SCORE
7	7 - 7.6 = -0.6	-0.6 ÷ 0.52 = -1.1538	-1.15
8	8 - 7.6 = 0.4	0.4 ÷ 0.52 = 0.7692	0.77
9	9 - 7.6 = 1.4	1.4 ÷ 0.52 = 2.6923	2.69
9	9 - 7.6 = 1.4	1.4 ÷ 0.52 = 2.6923	2.69
6	6 - 7.6 = -1.6	-1.6 ÷ 0.52 = -3.0769	-3.08
10	10 - 7.6 = 2.4	2.4 ÷ 0.52 = 2.7586	-2.76
7	7 - 7.6 = -0.6	-0.6 ÷ 0.52 = -1.1538	-1.15
5	5 - 7.6 = -2.6	-2.6 ÷ 0.52 = -4.6154	-4.62
9	9 - 7.6 = 1.4	1.4 ÷ 0.52 = 2.6923	2.69
6	6 - 7.6 = -1.6	-1.6 ÷ 0.52 = -3.0769	-3.08
$\sum \overline{X}$ = 76 n = 10			to 2 decimal places

S_x÷$\sqrt{n}$ = 0.52

Figure 42b: Assignment feedback - The z-scores.

UNIT 4C: ANALYZING INTERVAL DATA - THE t-test

PURPOSE:

The t-test is one of the most common statistical measures used in research. The t-test is used with interval (normal) data, when comparing the mean for one set of data either with a hypothesized mean or with the mean for a second set of data. The z-test works best with large samples; the t-test works with any size sample but is particularly helpful for smaller datasets. The flow chart shows the three paths to using the t-test. In all three situations, the test is based on an average or mean.

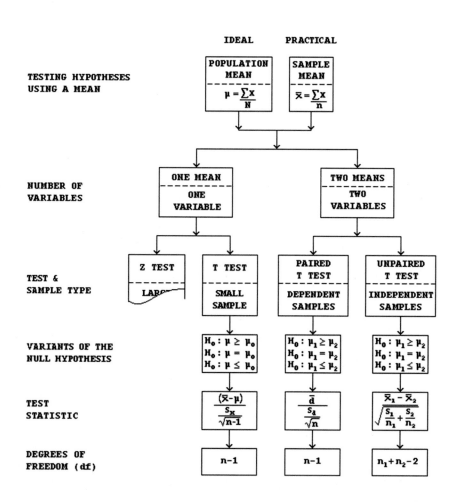

Figure 43: The three paths for doing the t-test.

Testing the hypothesis ideally uses the population mean. Unfortunately, the population mean is seldom known, so what is referred to as an hypothesized mean must be used for the path where there is only a single variable.

OBJECTIVES:

- Determine when it is appropriate to use the t-test.

- Given several research situations determine if the situation should be classified as:

  ◦ One group, with a group mean compared with a population mean.

  ◦ Two groups unpaired (independent samples), comparing the mean from one group with the mean from a totally different group.

  ◦ Two groups paired (dependent samples), comparing a mean for one group's dataset with the same group's mean from a different time or a different subject.

INTRODUCTION:

The ideal way to conduct research is to obtain data from the whole of the population in which we are interested or to be able to make a comparison between the population mean, μ or **sigma**, and the sample mean, $\overline{X}$. In most cases, trying to obtain data from the whole population would be impractical and, in real life, we rarely know exactly what the value of μ is for the population. There are situations where a population **is** known. For instance, enough I.Q. tests have been conducted all over the world that it has been established that the average I.Q. is 100 points. It is unlikely that average could change in the future.

The ACT is another situation where enough tests have been done to define the population mean. But this population is not the whole population of high school seniors in the U.S. The population is probably "all senior high school students who wish to go to college." This population might be stretched to include junior high school students who had taken advanced placement studies. Can this average or mean change? Possibly, but it might depend on whether the population changed to, say, "only high school seniors who wish to go to college" or "only high school seniors who had an I.Q. of more than 100 points." In addition, the ACT is an annual test, and the test to be taken varies. Possibly one ACT test may be more difficult than another which would affect the results. This suggests that the type of population, the makeup of the students taking the ACT, and the test taken all may affect the results of the ACT.

T-TEST TO ANALYZE THE VARIABILITY BETWEEN GROUPS:

It is important to realize that in making the comparison of the means in any situation, we must not just look at the two values and make our inferences directly from these values that the means are or are not equal. To make a valid comparison of the means we obtain, we need to take into account the variability of the datasets. To aid us in making this determination, we can use the analytical tool called the **t-test** together with the parameter called **degrees of freedom (df)** discussed in Unit 4C-1.

Another source of variability is in the sample we select from the population. Suppose we obtained a random sample of 200 students in order to find out how many Math courses they completed during the previous year. Let us say we determined the sample mean is 5 Math courses. But we cannot be certain that the mean for the population of all students during that year would also be 5 Math courses. Since the university keeps records on both a semester and annual basis, we could find out. If the selection of the subjects of the study is truly random, it is likely, but not certain, that our sample is representative of the population.

But suppose that the 200 students are **not** representative of all students. For example, it could be that one third of the entire population of students that year majored in science, and, it so happens, science majors make up only a quarter of our sample. Even if the proportion of science students in the sample is close to one third, there is also a possibility that our sample might contain a higher than normal proportion of students who have not yet taken **any** Math courses.

This example shows you that even when we do our best to select a representative group for our study, it is possible that we do not have a good representation of the total population. That is why in some studies, the researcher draws several random samples from the same population. If you only collect a single sample, then the t-test requires you to use an actual population mean or to hypothesize one.

A HYPOTHESIZED MEAN:

What is a hypothesized mean? It is a mean predicted or assumed for the population. How do we arrive at it? We may be able to get one from past results of many tests. For example, IQ has been tested so many times that the average IQ is known to be 100. Of course, if we are working with graduate students, the data may be skewed upwards because it is possible that only people with higher IQs attend graduate school. If that became the hypothesis, we would compare the average I.Q. (μ of 100) with the average I.Q. of the current graduate enrollment. Since that could be a large group, we may sample the graduate population in order to make the comparison.

But there are situations when there is insufficient data to provide us with a real population mean. Then we have to hypothesize it. By comparing the sample mean with the hypothesized mean, we can determine if they are similar or different and whether the difference is significant.

So, how do we get a hypothesized mean? One way is to take data from previous years to determine the population mean. For example, if we have access to data in the university's records, we could go back, say for the previous five years. Then we could calculate the number of math courses taken by the students for each year. We could compare the current year with a particular year or we could take an average of five years and use that for the hypothesized mean.

Frequently, the hypothesized mean is based on data from an earlier time or on an accumulation of data. For instance, suppose university administrators wish to see trends across several years, such as the types of freshman registering, or the choices for majors among sophomores. Comparing the current intake of freshman with data from past years allows for a hypothesized mean and the null hypothesis, H_0, would be that this year's intake would not be different from previous years. The alternative hypothesis, H_a, could state that the current intake would be significantly different from the intake of a set time. In a sense, we would be treating the current population of freshman as a sample and comparing it with the whole population of freshman at the university, for a specific year or a specific period of time, often 5 years or 10 years. If we chose a different period of time, say 50 years, we would increase the probability that the current group of freshmen would be significantly different from those of fifty years ago.

Universities keep records for all students who have graduated. Over a period of time, say the past five years, we can come up with a mean for the University's population and use this in our study. But we need to be careful in our selection of the population. For example, the population of students that attended the university fifty years ago are likely to be very different from the current population because of the historic changes that have affected the entire population for the U.S. Great-Great-Grandma is likely to be a very different person from her Great-Great-Granddaughter.

WHEN TO USE THE t-TEST:

The t-test is used whenever we are trying to draw an inference from one or two samples. The test enables us to determine whether the two means we are comparing are significantly different (the alternative hypothesis). One of the means is from our random sample the other is either the mean hypothesized for the population, or the mean from a second sample. We can use the t-test provided we are dealing with either one or two groups, the data we are collecting is in the interval scale and we have a mean for each group being considered:

1. The mean for data obtained from **one random sample** is significantly different from some **predicted or hypothesized mean**.

2. The means for data obtained from **two different and independent samples**, such as samples from the populations of two different schools, are significantly different.

3. The mean for data obtained from the same sample at **two different times therefore dependent** are or are not significantly different.

There are two important aspects of working with t-tests. One is degrees of freedom discussed in UNIT 4C-1. The other has to do with probability. In connection with the t-distribution (the probability curve), a cumulative probability is shown by an inequality, that is, it refers to the probability that a t-value or a sample mean will be less than or equal to a specified value. Suppose that we sample 100 first-graders. If we ask about the probability that the average first grader weighs **exactly** 70 pounds, we are asking about a **simple probability**. But if we ask about the probability that **average weight** is less than or equal to 70 pounds, we are really asking about a sum of probabilities, i.e., the probability that the average weight is exactly 70 pounds plus the probability that it is 69 pounds plus the probability that it is 68 pounds, etc. Thus, we are asking about a **cumulative probability**.

WHEN NOT TO USE THE t-TEST:

While the t-test is a useful tool, there are times when the t-test is inappropriate. This test should NOT be used in the following circumstances:

<u>THE DATA IS ORDINAL OR NOMINAL.</u>

For example, in a questionnaire, where a 5-point scale is often used (referred to as a Likert scale), where subjects are given a statement and asked for their preference, such as, Strongly Agree, Agree, No Preference, Disagree, Strongly Disagree. Each response then receives an ordinal value (ranking of 1-5) based on the subject's selection. In this case the **ranking** of the scores is the important data. The t-test is designed to work with interval data not with ranked data.

WHEN THERE ARE MORE THAN TWO GROUPS TO BE COMPARED.

The **AN**alysis **O**f **VA**riance (**ANOVA**, a computerized program) is more appropriate for comparing the means in such cases. When you work on your thesis or dissertation you are likely to have more than two variables. In that case, you may work with a statistics graduate student who will act as your "expert" and he will be - that's why he or she is studying statistics.

THREE SITUATIONS FOR USING THE t-TEST:

There are three different t-tests depending on what type of situation your data describes. All three require the analysis of means.

The first is when you have one mean with a single variable, that is, a single small sample.

The second is when you have two samples that are dependent on each other. This compares the same group across different times, such as a midterm and a final exam on the same topic; or across two different topics such as Math and English; because the same group of subjects are involved, this is referred to as a **paired group**.

The third is when the subjects of your study are in two groups that a totally independent of each other, such as certain students at Elementary School A compared with certain students at Elementary School B. The topic of the study is the same but the subjects of group **A** are totally different from the subjects of **B**.

SINGLE GROUP	TWO GROUP DEPENDENT (PAIRED)	TWO GROUP INDEPENDENT
Compare the mean of a data set from a single group with a population or hypothesized mean.	Compare the means of two data sets from the same subjects at different times or for different topics.	Compare the means of two data sets from two independent groups on the same topic.

Figure 44: Three criteria for using the t-test and the research question for each.

ASSIGNMENT 4C: USING THE T-TEST.

Classify the following research questions into "two group independent", "two group paired", or "one group" situations.

_____ 1.   It was hypothesized that the acid content of paper in books published by Holt, Rinehart & Winston would be 0.6. A sample of 100 books published by the company was tested for the acid content of the paper.

_____ 2.   It was hypothesized that the first grade students at Washington Elementary would perform as well in mathematics as first grade students at Jefferson Elementary. A sample of 80 first graders from each school were tested for their skills in the core math curriculum at the end of the first semester.

_____ 3.   It was hypothesized that the math scores of high school students at Mountainland High would remain constant over a period of three years. The math skills of a sample of 120 students was tested at the end of 9th grade, and the same sample was tested again at the end of 11th grade.

_____ 4.   A comparison was made between 100 seniors majoring in nuclear engineering, and 100 seniors majoring in Psychology, with reference to the mean number of books checked out of the university library by each group.

_____ 5.   A comparison was made between the mean number of books checked out by the 100 students in the fall of their Freshman year, and the mean number of books checked out by the same students in the Fall of their Sophomore year.

_____ 6.   It was hypothesized that the mean number of students using the university bookstore between 2.00 p.m. and 8.00 p.m. would be 105. Data was obtained on the actual number of students using the bookstore during those hours in the Fall of their Sophomore year.

ASSIGNMENT 4C: USING THE T-TEST – feedback.

The correct response to each question and an explanation of those responses are included below.

Example 1:   One Group.

We are comparing the mean for one group of books with a hypothesized mean.

Example 2:   Two group independent sample.

Comparing two samples from two different schools, that is two independent groups.

Example 3:   Two group paired data.

The data is obtained from the same sample of students at two different times.

Example 4:   Two group independent sample.

Two independent groups are compared with reference to the mean number of books each group checked out.

Example 5:   Two groups paired data.

The data was obtained for the same group at two different times.

Example 6:   One group.

The mean of actual data was compared with the hypothesized mean.

UNIT 4C-1: DEGREES OF FREEDOM

PURPOSE:

Degrees of freedom refer to a measure that helps to increase the accuracy of the statistical analysis tests. The z-test does not need to use degrees of freedom, known as *df.* The rest of the tests that will be discussed do use degrees of freedom.

OBJECTIVES:

- Define the term "Degrees of Freedom" and describe how you determine the **df** for a single-tail and for a two-tail distribution.

- Use the URL to determine the critical value for a specific **df**:

  http://www.tutor-homework.com/statistics_tables/statistics_tables.html

INTRODUCTION:

The t-test and other calculations require the use of **degrees of freedom (df)** to increase the accuracy of your calculation, that is, to make it closer to the population parameter you are interested in. The following information was extracted from the internet concerning degrees of freedom. The information seems to be particularly helpful.

Dr. Dallal's definition of degrees of freedom[4]: "At the moment, I'm inclined to define **degrees of freedom as a way of keeping score.** A data set contains a number of observations, say, *n*. They constitute *n* individual pieces of information. These pieces of information can be used either to estimate **parameters** or **variability**. *In general, each item being estimated costs one degree of freedom. The remaining degrees of freedom are used to estimate variability. All we have to do is count properly.*" [Bold & Italic emphasis added.] The following is based on his article:

HOW DO YOU COUNT DEGREES OF FREEDOM?

Degrees of freedom can be described as the number of scores that are free to vary. For example, suppose you tossed three dice. The total score for the three dice adds up to 12. If you rolled a 3 on the first die and a 5 on the second, then you know that the third die must be a 4, otherwise the total would not add up to 12. In this example, 2 die are free to vary while the third is not. Therefore, there are 2 degrees of freedom. Notice that the degrees of freedom are **one less** than the number of observations made, in this case three. If the sample size were 20, there would be 20 observations and the degrees of freedom would be 20 minus 1 or 19. This means that there are 19 degrees of freedom. The 20th observation is not free to vary.

Consider tossing a die three times. Suppose you need to make a total of 13. The possible outcomes of the first two tosses include throwing a six each time. If you do, the total at the end of the second throw is 12. That means your next throw can only be a one, the freedom to throw any other amount has been lost. Suppose your first two tosses produce a total of 8, then to make a total of 13, your next throw must be a five. Again, based on the total of the first two tosses, your freedom to toss anything other than five has been lost. In either toss you have two (2) degrees of freedom which is one less than the number of observations or 3 minus 1 or 3-1 = 2.

Degrees of freedom are calculated based on the number of observations in your dataset. There are ways of determining the degrees of freedom for any appropriate test.

---

4 "Degrees of Freedom" by Gerard E. Dallal, Ph.D.
   URL: http://www.jerrydallal.com/LHSP/dof.htm]
   Used with permission from Gerard E. Dallal.

EXAMPLES OF DEGREES OF FREEDOM:

1. If you have a single sample for your dataset with **n** observations, there is only one parameter (the mean) that needs to be estimated. That leaves **n-1** degrees of freedom for estimating variability, **P**.

2. Suppose you have two different samples in your dataset: Each set with variables X_1 and X_2, has its own number of observations, which may or may not be equal, written as: n_1 and n_2; there are a total of $n_1 + n_2$ observations. With two datasets, there are two means (parameters) to be estimated, leaving $n_1 + n_2 - 2$ degrees of freedom for estimating variability (**P**). Another way to consider it is: dataset X_1 has $n_1 - 1$ degrees of freedom and Dataset X_2 has $n_2 - 1$ degrees of freedom. Together the two sets have ($n_1 - 1 + n_2 - 1$) degrees of freedom or $n_1 + n_2 - 2$.

3. The internet has numerous calculators for you to use in determining the t-distribution. You will need to know the df in order to uses these.

The table below is a partial df-distribution table which shows degrees of freedom in the left column and the percentage for the probability across the top of the table.

Example: suppose you have calculated the t-value as 2.55 and there are 8 observations:

df \ P	90%	95%	97.5%	99%	99.5%	99.9%
1	3.078	6.314	2.706	31.821	63.657	318.313
2	1.886	2.920	4.303	6.965	9.925	22.327
3	1.638	2.353	3.182	4.541	5.841	10.215
4	1.533	2.132	2.776	3.747	4.604	7.173
5	1.476	2.015	2.571	3.365	4.032	5.893
6	1.440	1.943	2.447	3.143	3.707	5.208
7	1.415	1.895	2.365	2.998	3.499	4.782
8	1.397	1.860	2.306	2.896	3.355	4.499
9	1.383	1.833	2.262	2.821	3.250	4.296
10	1.372	1.812	2.228	2.764	3.169	4.143
11	1.363	1.796	2.201	2.718	3.106	4.024
12	1.356	1.782	2.179	2.681	3.055	3.929
13	1.350	1.771	2.160	2.650	3.012	3.852
14	1.345	1.761	2.145	2.624	2.977	3.787
15	1.341	1.753	2.131	2.602	2.947	3.733
16	1.337	1.746	2.120	2.583	2.921	3.686
17	1.333	1.740	2.110	2.567	2.898	3.646
18	1.330	1.734	2.101	2.552	2.878	3.610
19	1.328	1.729	2.093	2.539	2.861	3.579
20	1.325	1.725	2.086	2.528	2.845	3.552
21	1.323	1.721	2.080	2.518	2.831	3.527
22	1.321	1.717	2.074	2.508	2.819	3.505
23	1.319	1.714	2.069	2.500	2.807	3.485
24	1.318	1.711	2.064	2.492	2.797	3.467
25	1.316	1.708	2.060	2.485	2.787	3.450
					2.779	3.435

1. Look at the left column for degrees of freedom: 8-1 = 7 so df = 7.

2. Look along the columns for the t-value, in this example, 2.55. The value 2.55 lies between 2.365 in the 97.5% column and 2.998 in the 99% column.

3. You would write it as 97.5 < P < 99.0 which reads: "P lies between 97.5 and 99." This is as close as you can get using this particular table.

Figure 45: Degrees of freedom chart shows lack of precision.

ON-LINE CALCULATORS:

[Note: The graphics here were obtained from the internet screen by using the program: **SnagIt.**]

Using the URL: http://stattrek.com/Tables/t.aspx has a calculator but there is no graph available. The title is "T Distribution Calculator: Online statistical table.

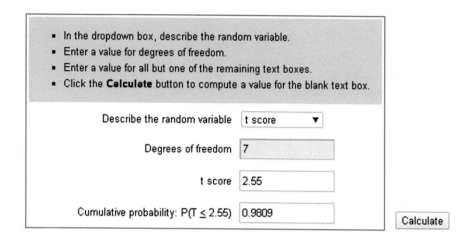

Figure 46: The Stattrek site gets the same results but using another method.

1. The top box asks you to "Describe the random variable." In this case it is the "T score" [NOTE: the alternative random variable is "Sample mean"].

2. Type in the degrees of freedom (7).

3. Type in the **T score** (2.55) and press the "Calculate" button to get a cumulative probability of 0.9809.

There is another site on the Internet which has the URL:

http://www.tutor-homework.com/statistics_tables/statistics_tables.html

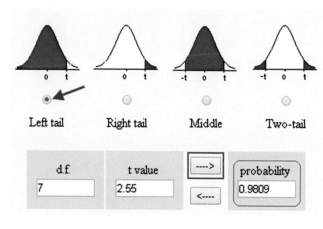

Figure 47a: Left tail chosen, gives 0.9809.

The statistics table (second of four) found at this site is called the ***student's t-distribution.*** The name comes from a study during which the researcher, in an effort to keep anonymity, called himself "student" with the result that this distribution was called a student's t-distribution.

NOTE: There are four distributions shown at this site. The default setting is the left tail. To see how this works: With the default setting, write 7 degrees of freedom in the "df" box. Then, just for an example, write 2.55 in the "t-value" box. Then press the right-pointing arrow to initiate the calculation. The probability is calculated as 0.9808.

In other words, the probability of having a left-tailed t-value of 2.55 at 7 df is 98%. This is even more significant than 95%.

Compare the left tail picture with the right-tail picture. The red portion of the latter is much smaller.

Select the dot for the right tail then press the right-pointing arrow to initiate the calculation. This time the probability is only 0.0191. In other words, the likelihood of having a right-tailed t-value of 2.55 at 7 df is very low.

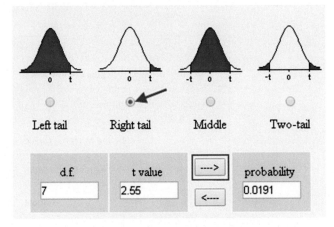

Figure 47b: Right tail chosen, gives a probability of 0.0191.

This site's t-distribution calculator is probably the best one to use when you are working on the t-test.

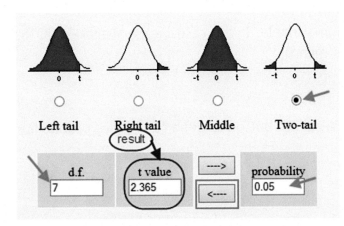

Figure 47c: Two tail chosen, gives t-value of 2.365.

You can also use this site to calculate the critical t value. This time you select the two-tail dot, insert the df (7)and the desired probability (0.05) and press the lower direction arrow to get 2.365 as the critical t-value.

This internet site will be used in the examples following (UNIT 4C-2, etc.)

EXAMPLE USING THE t-TABLE CHART:

The most commonly used hand chart is this one. It is better than the one in Figure 45, as it shows both one- and two-tailed data. Two-tailed information is offset one column to the right.

t Table

one-tail	0.50	0.25	0.20	0.15	0.10	0.05	0.025	0.01	0.005	0.001	0.0005
two-tails	1.00	0.50	0.40	0.30	0.20	0.10	0.05	0.02	0.01	0.002	0.001
df 1	0.000	1.000	1.376	1.963	3.078	6.314	12.71	31.82	63.66	318.31	636.62
2	0.000	0.816	1.061	1.386	1.886	2.920	4.303	6.965	9.925	22.327	31.599
3	0.000	0.765	0.978	1.250	1.638	2.353	3.182	4.541	5.841	10.215	12.924
4	0.000	0.741	0.941	1.190	1.533	2.132	2.776	3.747	4.604	7.173	8.610
5	0.000	0.727	0.920	1.156	1.476	2.015	2.571	3.365	4.032	5.893	6.869
6	0.000	0.718	0.906	1.134	1.440	1.943	2.447	3.143	3.707	5.208	5.959
7	0.000	0.711	0.896	1.119	1.415	1.895	2.365	2.998	3.499	4.785	5.408
8	0.000	0.706	0.889	1.108	1.397	1.860	2.306	2.896	3.355	4.501	5.041
9	0.000	0.703	0.883	1.100	1.383	1.833	2.262	2.821	3.250	4.297	4.781
10	0.000	0.700	0.879	1.093	1.372	1.812	2.228	2.764	3.169	4.144	4.587
11	0.000	0.697	0.876	1.088	1.363	1.796	2.201	2.718	3.106	4.025	4.437
12	0.000	0.695	0.873	1.083	1.356	1.782	2.179	2.681	3.055	3.930	4.318
13	0.000	0.694	0.870	1.079	1.350	1.771	2.160	2.650	3.012	3.852	4.221
14	0.000	0.692	0.868	1.076	1.345	1.761	2.145	2.624	2.977	3.787	4.140
15	0.000	0.691	0.866	1.074	1.341	1.753	2.131	2.602	2.947	3.733	4.073
16	0.000	0.690	0.865	1.071	1.337	1.746	2.120	2.583	2.921	3.686	4.015
17	0.000	0.689	0.863	1.069	1.333	1.740	2.110	2.567	2.898	3.646	3.965
18	0.000	0.688	0.862	1.067	1.330	1.734	2.101	2.552	2.878	3.610	3.922
19	0.000	0.688	0.861	1.066	1.328	1.729	2.093	2.539	2.861	3.579	3.883
	0.697	0.860	1.064	1.325	1.725	2.086	2.528	2.845	3.552	3.850

Figure 48; The t-table chart for one- and two-tailed distributions.

t Table

one-tail	0.50	0.25	0.20	0.15	0.10	0.05	0.025	0.01	0.005	0.001	0.0005
two-tails	1.00	0.50	0.40	0.30	0.20	0.10	0.05	0.025	0.01	0.002	0.001
df											
1	0.000	1.000	1.376	1.963	3.078	6.314	12.71	31.82	63.66	318.31	636.62
2	0.000	0.816	1.061	1.386	1.886	2.920	4.303	6.965	9.925	22.327	31.599
3	0.000	0.765	0.978	1.250	1.638	2.353	3.182	4.541	5.841	10.215	12.924
4	0.000	0.741	0.941	1.190	1.533	2.132	2.776	3.747	4.604	7.173	8.610
5	0.000	0.727	0.920	1.156	1.476	2.015	2.571	3.365	4.032	5.893	6.869
6	0.000	0.718	0.906	1.134	1.440	1.943	2.447	3.143	3.707	5.208	5.959
7	0.000	0.711	0.896	1.119	1.415	1.895	2.365	2.998	3.499	4.785	5.408
8	0.000	0.706	0.889	1.108	1.397	1.860	2.306	2.896	3.355	4.501	5.041
9	0.000	0.703	0.883	1.100	1.383	1.833	2.262	2.821	3.250	4.297	4.781
10	0.000	0.700	0.879	1.093	1.372	1.812	2.228	2.764	3.169	4.144	4.587
11	0.000	0.697	0.876	1.088	1.363	1.796	2.201	2.718	3.106	4.025	4.437
12	0.000	0.695	0.873	1.083	1.356	1.782	2.179	2.681	3.055	3.930	4.318
13	0.000	0.694	0.870	1.079	1.350	1.771	2.160	2.650	3.012	3.852	4.221
14	0.000	0.692	0.868	1.076	1.345	1.761	2.145	2.624	2.977	3.787	4.140
15	0.000	0.691	0.866	1.074	1.341	1.753	2.131	2.602	2.947	3.733	4.073
16	0.000	0.690	0.865	1.071	1.337	1.746	2.120	2.583	2.921	3.686	4.015
17	0.000	0.689	0.863	1.069	1.333	1.740	2.110	2.567	2.898	3.646	3.965
18	0.000	0.688	0.862	1.067	1.330	1.734	2.101	2.552	2.878	3.610	3.922
19	0.000	0.688	0.861	1.066	1.328	1.729	2.098	2.539	2.861	3.579	3.883
20	0.000	0.687	0.860	1.064	1.325	1.725	2.086	2.528	2.845	3.552	3.850
21	0.000	0.686	0.859	1.063	1.323	1.721	2.080	2.518	2.831	3.527	3.819
22	0.000	0.686	0.858	1.061	1.321	1.717	2.074	2.508	2.819	3.505	3.792
23	0.000	0.685	0.858	1.060	1.319	1.714	2.069	2.500	2.807	3.485	3.768
24	0.000	0.685	0.857	1.059	1.318	1.711	2.064	2.492	2.797	3.467	3.745
25	0.000	0.684	0.856	1.058	1.316	1.708	2.060	2.485	2.787	3.450	3.725
26	0.000	0.684	0.856	1.053	1.315	1.706	2.056	2.479	2.779	3.435	3.707
27	0.000	0.684	0.855	1.057	1.314	1.703	2.052	2.473	2.771	3.421	3.690
28	0.000	0.683	0.855	1.056	1.313	1.701	2.043	2.467	2.763	3.408	3.674
29	0.000	0.683	0.854	1.055	1.311	1.699	2.045	2.462	2.756	3.396	3.659
30	0.000	0.683	0.854	1.055	1.310	1.697	2.042	2.457	2.750	3.385	3.646
40	0.000	0.681	0.851	1.050	1.303	1.684	2.021	2.423	2.704	3.307	3.551
60	0.000	0.679	0.848	1.045	1.296	1.671	2.000	2.390	2.660	3.232	3.460
80	0.000	0.678	0.846	1.043	1.292	1.664	1.990	2.374	2.639	3.195	3.416
100	0.000	0.677	0.845	1.042	1.290	1.660	1.984	2.364	2.626	3.174	3.390
1000	0.000	0.675	0.842	1.037	1.282	1.646	1.962	2.330	2.531	3.093	3.300
	0%	50%	60%	70%	80%	90%	95%	98%	99%	99.8%	99.9%

Confidence Level

Figure 49: critical t when there are 24 df and 1- and 2-tail distributions.

Let's look at an example and the fact that the information read from the chart gives us the *critical t-value* that enables us to determine if H_0 may be rejected or not.

Given: There are 24 **df** and a significance level of 0.01, use the table to determine the critical value for a one-tail (red marker) and a two-tail (green marker) t-test.

With a one tail distribution, *critical t* at 24 df and 0.01 significance is 2.492.

With a two-tail distribution, *critical t* at 24 df and 0.01 significance is 2.797.

UNIT 4C-2: THE t-test CRITICAL VALUE & CALCULATED VALUE

PURPOSE:

The t-test is one of the most common statistical measures used in research. The t-test is used with interval (normal) data, when comparing the mean for one set of data either with a hypothesized mean or with the mean for a second set of data. While the z-test works best with large samples, the t-test works with any size sample but is particularly helpful for smaller datasets. The t-test is used with interval data, when comparing the mean for one set of data either with a hypothesized mean or with the mean for a second set of data.

The flow chart shows the three paths to using the t-test. In all three situations, the test is based on an average or mean. Testing the hypothesis ideally uses the population mean. Unfortunately, the population mean is seldom known, so what is referred to as a **hypothesized mean** must be used for the path where only a single variable is used.

OBJECTIVES:

Determine when it is appropriate to use the t-test.

- Given some research situations determine if the situation should be classified as:

    ◦ One group, with a group mean compared with a population mean.

    ◦ Two groups unpaired (independent samples), comparing the mean from one group with the mean from a totally different group.

    ◦ Two groups paired (dependent samples), comparing a mean for one group's dataset with the same group's mean from a different time or a different subject.

- Perform each of the three t-tests and interpret the results.

INTRODUCTION:

When performing any of the t-tests, there are two values that need to be considered. First, the **critical value** found by using a t-table, with the combination of degrees of freedom (**df**) and the level of significance (**α**). The "shorthand" for showing you are looking at a critical value is:

Critical t$_{(df)(\alpha)}$ Example: **Critical t**$_{(14)(0.01)}$ = **2.977**

The critical value is the measure of whether the calculated value is significant. The **calculated value** is found by using one of the formulas shown in the t-test flow chart and determined by the type of t-test needed:

REVIEW (RELEVANT TERMINOLOGY):

STATISTICAL SIGNIFICANCE:

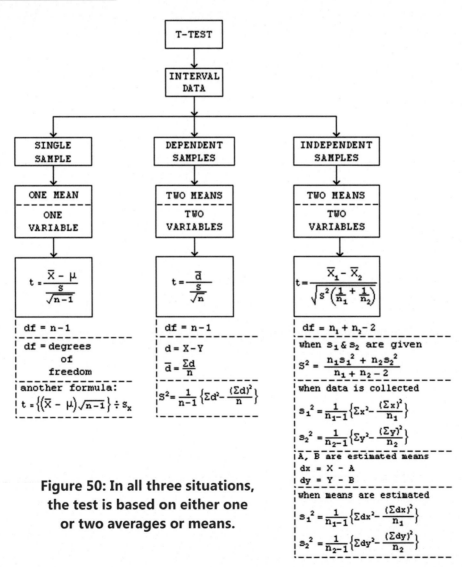

Figure 50: In all three situations, the test is based on either one or two averages or means.

When you have collected your data, you use that data and the appropriate statistical test (t-, z-, Spearman Correlation, Pearson Correlation, or Chi-Square) to determine if the results of your study are statistically significant, that is, whether the results are due to chance or not. Before you begin the analysis you determine the level of significance you want to use: 0.05 represents 5 possibilities in 100 are by chance, therefore there are 95 chances that your data is significant; 0.01 represents 1 chance in 100, therefore, the significance is greater.

NULL HYPOTHESIS:

The null hypothesis is usually the hypothesis that your sample observations result only from chance. That means that **H_0 is always written as an equality.** Note that each statement contains the words: "***equal to***" which is why all three are referred to as "equalities." There are three possible equalities:

1. ***less than or equal to*** (≤) [giving a range of values]

2. ***greater than or equal to*** (≥) [also giving a range of values]

3. is exactly ***equal to*** (=) [giving one specific value]

TAILS:

By mathematical convention, ≤ refers to values to the **left of** the reference quantity and ≥ refers to values to the **right of** the reference quantity. The reference quantity is often the mean of your data. The direction of the inequality depends on whether the hypothesis is right-tailed, left-tailed or two tailed. H_0 in each case is shown by the red areas of the curves.

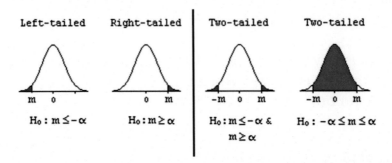

Figure 51: Based on the null and alternative hypotheses for one-tailed and two-tailed curves, the red areas show the regions of rejection.

Note that for the one-tailed curves, there is only a single area of rejection. For the two-tailed curves, depending on the null hypothesis, there are either two areas of rejection or a single area of rejection (this distribution is sometimes referred to as middle tailed). The areas of rejection are totally dependent on the stated null hypothesis. Note also, that the values, α and −α, are variables dependent on the data collected.

What is sometimes referred to as a **"special case"** is where the null hypothesis has only a single value and the areas of rejection (red) are on both sides of that value:

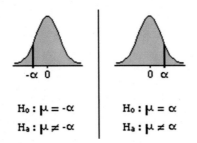

Figure 52: Both curves show two-tailed rejection regions.

CRITICAL & CALCULATED VALUES:

There are two values you use to determine significance of your data: **Critical** and **Calculated values**. The *critical value* is determined from a table associated with the type of test you are using combined with the rejection region determined by your hypothesis, and the *calculated value* is found by using that test to analyze your data. Since your hypothesis is unique to your study, so too are the left-, right-, and two-tailed regions unique to your hypothesis. NOTE:

left- and right-tailed distributions are another name for a one-tailed distribution. The direction of the tail determines the rejection region.

RULE 1 - LEFT-TAILED REJECTION REGION:

When you have a *left-tailed rejection region,* fail to reject the null hypothesis if the *calculated value < critical value.*

RULE 2 - RIGHT-TAILED REJECTION REGION:

When you have a *right-tailed rejection region,* fail to reject the null hypothesis if the *calculated value > critical value.*

RULE 3 - TWO-TAILED REJECTION REGION:

When you have a *two-tailed rejection region,* fail to reject the null hypothesis if the |*calculated value*| > critical value.

Reminder: the symbol "|calculated *value*|" means that the **absolute value** is to be considered (that is, ignore any positive or negative sign).

Examples: $|-3| \rightarrow$ use 3.

$|+3| \rightarrow$ use 3.

$|\pm3| \rightarrow$ use 3.

The above rules apply whether you are doing a t-test, a z-test, a Pearson or Spearman correlation test or a Chi-Square test. In each type of test, a different formula and a different table is used to calculate the value of t, the test statistic, and to determine the critical value.

TESTING HYPOTHESES:

There is a different formula to use with each situation. One t-test is used when dealing with a single group (that is, group-to-population) and the other is used when dealing with either of the two group circumstances, that is, two-group-dependent or two-group-independent.

ALTERNATIVE FORMULAS:

When you find sites on the internet dealing with statistics, you may find that a different formula is used for the same test. Ignore any differences and use the ones given in this unit. Generally, they are the same formula just in a different form. For example, the basic t-test formula for testing a small independent sample with a single mean is to the left:

$$t = \frac{(\bar{X} - \mu)}{\frac{s_x}{\sqrt{n-1}}}$$

Here is the formula given in the three situations flow chart (Figure 52) for conducting a t-test (note that **n** may be used instead of **n-1**). Let's call this formula (1) where the denominator is a fraction.

[Note: One mnemonic for remembering the position of the numerator is that 'u' says 'up' and for the **d**enominator, 'd' says 'down'.]

The denominator of formula (1) is a fraction, that is, S_x divided by $\sqrt{(n-1)}$. To get rid of the fraction in the denominator you would need to multiply the denominator by $\sqrt{(n-1)}$. The rules for fractions say that if you multiply the denominator by anything, you must multiply the numerator by the same thing. That means: multiply the numerator and denominator, by another name for one, in this case:

$$\frac{\sqrt{n-1}}{\sqrt{n-1}} = 1$$

This gives us:

$$t = \frac{(\bar{X} - \mu)}{\dfrac{S_x}{\sqrt{n-1}}} * \frac{\sqrt{n-1}}{\sqrt{n-1}} \qquad \text{Where * means 'multiply by'}$$

Multiply to get Formula 2:

$$t = \frac{(\bar{X} - \mu) * \sqrt{n-1}}{S_x}$$

Formula 2 can be written with a division sign as formula 3:

$$t = [(\bar{X} - \mu) * \sqrt{n}] \div S_x \qquad \text{Note: both formula 2 \& 3 are equivalent to the original formula.}$$

UNIT 4C-3. THE GROUP-TO-POPULATION t-test

PURPOSE:

When only a single set of data is collected you need to compare the mean of the collected data with the population mean. But generally it is not possible to get a real population mean. In that case you will need to estimate that mean, that is, you must determine a hypothesized population mean. We may wish to compare the mean for this data with a hypothesized mean. The hypothesized mean can be from data from earlier studies, an expected mean based on other data, or a mean based on experience.

For example, we may hypothesize that the mean age of all the students in ED Psych 508 is 25 years. To test this hypothesis, we obtain the actual ages of a

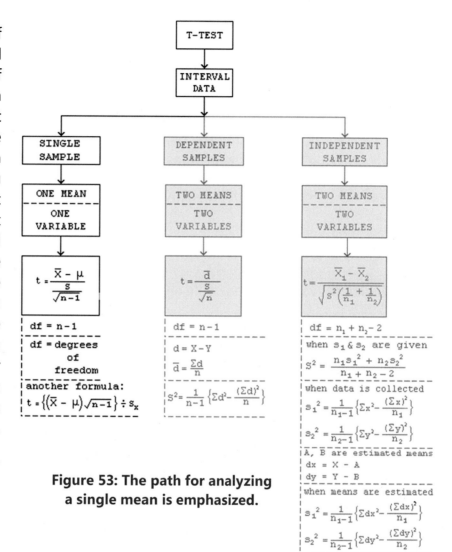

Figure 53: The path for analyzing a single mean is emphasized.

sample of students currently taking ED Psych 508 and conduct a t-test to see if the mean age of the students is significantly different from the hypothesized mean of 25 years.

OBJECTIVES:

- Determine the hypothesized mean, when you have a single set of data, that is, a single variable.

- Perform a group-to-population t-test, that is, compare the variable mean with a hypothesized mean and interpret the results.

CALCULATING THE t-value FROM SCRATCH:

In the example given above, the mean and standard deviation were given. But the data you collect will require you to calculate both; to do that you need an algorithm. Since our aim is to reject the null hypothesis, the algorithm includes finding evidence that the null hypothesis is false.

ALGORITHM FOR THE ONE GROUP t-TEST:

1. Calculate the actual mean for the group $(\overline{X})$.

2. Subtract the hypothesized mean (μ) from the actual mean: $(\overline{X} - \mu)$.

3. Calculate the square root of one less than the total number of observations $\{\sqrt{(n-1)}\}$.

4. Multiply (symbol *****) the results of steps 2 and 3, $\{(\overline{X}-\mu)*\sqrt{(n-1)}\}$.

5. Calculate the standard deviation $\{\sqrt{\Sigma(X - \overline{X})^2}\}$ remember Σ = sum of.

6. Calculate the t-value by dividing the result of step 4 by the standard deviation $(\overline{X}-\mu)*\sqrt{(n-1)}/\{\sqrt{\Sigma(X - \overline{X})^2}\}$ where * means multiply.

7. Find the critical value on the t-table and compare with the calculated t-value.

8. Determine if H_0 can or cannot be rejected.

EXAMPLE 1:

student	score
1	81
2	76
3	51
4	65
5	56
6	73
7	85
8	91

Figure 57a: Unit 4C-3, Example 1, data.

The scores obtained by 8 students in a test are: 30, 22, 11, 51, 72, 83, 25, 90; the table below shows the ordered scores. We are told that the mean score (μ) for this group is **60** and the level of significance is **α = 0.05.** Does this set of scores produce a mean equal to the hypothesized mean.

THE HYPOTHESES:

$H_0: \overline{X} = \mu$

$H_a: \overline{X} \neq \mu$

ANALYZING THE HYPOTHESES:

There is only one value where the null hypothesis can be fulfilled, $\overline{X} = \mu$. If the mean holds any other value (alternate hypothesis) the null hypothesis must be rejected. Since these values lie on either side of the mean, we have a **two–tailed region of rejection**. Interpretation: if |calculated value| > critical value, cannot reject H_0.

STEP 1:

Calculate $\overline{X}$.

$$\overline{X} = \frac{81+76+51+65+70+85+91}{8}$$

$$= \frac{1434}{8}$$

$$= 72.25$$

STEP 2:

Subtract the hypothesized mean (μ) from the actual mean $(\overline{X} - \mu)$.

48 - 60 = -12

STEP 3:

Calculate the square root of one less than the total number of observations $\sqrt{(n-1)}$.

$\sqrt{(8-1)}$ = 2.65 (to 2 decimal places)

STEP 4:

Multiply the results of steps 2 and 3.

$(\overline{X}-\mu)*\sqrt{(n-1)}$

= -12 * 2.65

= -31.8

STEP 5:

Calculate S_x the standard deviation (see unit 3C-2)

The mean $\overline{X}$ = 48

- Subtract the mean $(\overline{X})$ from each observation.

- Square the value of $(X-\overline{X})$ for each observation.

- Find $\Sigma(X-\overline{X})^2$.

- Divide $\Sigma(X-\overline{X})^2$ by **(n-1)**,

- Then find S_x^2

student	score	$(X-\overline{x})$	$(X-\overline{x})^2$
1	81	81-72.25)= 8.25	68.06
2	76	76-72.25)= 3.75	14.06
3	51	51-72.25)=-21.25	451.56
4	65	65-72.25)= -7.25	52.56
5	56	56-72.25)=-16.25	264.06
6	73	73-72.25)= 0.75	0.56
7	85	85-72.25)= 12.75	162.56
8	91	91-72.25)= 18.75	351.56
			$\Sigma(X-\overline{x})^2 =$ 1364.98

Figure 54b: Unit 4C-3, Example 1, Step 5.

$$= \{\Sigma(X-\overline{X})^2 \div (n-1)\}$$

$$= \{6472 \div 7\}$$

$$= 924.57$$

$$S_x \quad = \sqrt{924.57}$$
$$= 30.41$$

STEP 6:

Calculate the t-value by dividing the result of **Step 4** by the standard deviation from **Step 5**.

From **STEP 4:**	$-31.8 \div 30.41 = -1.05$ (to two decimal places)
Given:	$\mu = 60$
	$\alpha = 0.05$
From **STEP 1:**	$\overline{X} = 48$
From **STEP 5:**	$S_x = 30.41$
From **STEP 2:**	$(\overline{X} - \mu) = -12$
From **STEP 6:**	**calculated value = -1.05**

The null and alternative hypotheses give us a two-tailed test with 7 degrees of freedom and a level of significance of 0.05:

t Table								
one-tail		0.50	0.25	0.20	0.15	0.10	0.05	0.025
two-tails		1.00	0.50	0.40	0.30	0.20	0.10	0.05
df	1	0.000	1.000	1.376	1.963	3.078	6.314	12.71
	2	0.000	0.816	1.061	1.386	1.886	2.920	4.303
	3	0.000	0.765	0.978	1.250	1.638	2.353	3.182
	4	0.000	0.741	0.941	1.190	1.533	2.132	2.776
	5	0.000	0.727	0.920	1.156	1.476	2.015	2.571
	6	0.000	0.718	0.906	1.134	1.440	1.943	2.447
	7	0.000	0.711	0.896	1.119	1.415	1.895	2.365
	8	0.000	0.706	0.889	1.108	1.397	1.860	2.306
	9	0.000	0.703	0.883	1.100	1.383	1.833	2.262
	10	0.000	0.700	0.879	1.093	1.372	1.812	2.228
	11	0.000	0.697	0.876	1.088	1.363	1.796	2.201
	12	0.000	0.695	0.873	1.083	1.356	1.782	2.179
	13	0.000	0.694	0.870	1.079	1.350	1.771	2.160

Figure 54c: Unit 4C-3, Example 1, Two-tails, df of 7 and P=0.05.

From the table, critical $t_{(df)(\alpha)}$ = critical $t_{(7)(0.05)}$ = 2.365.

Since the hypothesis is two-tailed, fail to reject the null hypothesis if |calculated value| > critical value.

|calculated t|= 1.05

Critical t = 2.365

1.05 < 2.365

The calculated value is less than the critical value so reject the null hypothesis.

CONCLUSION:

The mean of the sample set, 48, is significantly different from the hypothesized mean, 60.

ONE GROUP (GROUP-TO-POPULATION COMPARISONS):

Having selected a sample from a specific population, the group is studied to obtain the needed data: a set of observations in the form of numerical values. We may wish to compare the mean for this data with a hypothesized mean.

Note that the hypothesized mean can be derived from: data from earlier studies; or an expected mean based on other data; or a mean based on experience. An example: so many I.Q. tests have been made that if your research question concerns I.Q., the "hypothesized" mean is 100.

T-VALUE FOR ONE-GROUP SET OF OBSERVATIONS:

Example: five years ago, a college became a university and the administration wishes to know if the freshman intake has remained the same. The university has collected data for the years it was a college and the years since it has become a university.

Suppose the hypothesis is that the status of the institution does not affect the student intake. The "hypothesized" mean could be either for the five years before the change or for the last year the institution was a college. The observational mean could be either a mean for the latest year as a university or a mean for the five years since the conversion. In either case, the observational mean is being compared to a hypothesized mean.

Having selected a single sample from a specific population of freshman, the group is studied to obtain the needed data: a set of observations in the form of numerical values. We need to make a **group-to-population comparison.** That is, we wish to compare the observed values with the hypothesized value. This is a one-tailed distribution.

This is what the one-group t-test formula looks like:

student	score
1	81
2	76
3	51
4	65
5	56
6	73
7	85
8	91

$$t = \frac{(\bar{X} - \mu)}{\frac{S_x}{\sqrt{n-1}}}$$

Where $\bar{X}$ is the sample mean and μ is the hypothesized mean; n is the number of observations; and S_x is the standard deviation.

The distribution found using this formula is called the **student's distribution.** The name comes from a study which the researcher, in an effort to keep anonymity, called himself "student" with the result that the distribution was called a student's distribution. It is a way of estimating the mean of a population with a normal distribution when the sample is small and when the population standard deviation (sigma) is unknown, as is usually the case. The distribution is a normal curve. The tails of the curve are thicker than those in a standard distribution and reduce in size as the sample size increases. At a sample size of 30 or more, the t-distribution closely approaches the normal distribution.

EXAMPLE 2:

The same data is used but in a different context. Over a period of 8 hours the flow in a hydrothermal unit in gals per minute were recorded. The flow is normally 75 gals per minute. Do the sample results average 75 gallons per minute?

Hour	gals per min
1	81
2	76
3	51
4	65
5	56
6	73
7	85
8	91

Figure 55a: Unit 4C-3, Example 2, data.

81, 76, 51, 65, 56, 73, 85, 91.

The table shows the scores (they do not need to be sorted but you may prefer to).

Let us hypothesize that the mean score (μ) for this group is 75.

H$_0$: $\bar{X}$ = μ

H$_a$: $\bar{X}$ ≠ μ

This is a two-tailed distribution, therefore, if |calculated value| > critical value, fail to reject **H$_0$**. [NOTE: The desired outcome is to fail to reject H$_0$.]

STEP 1:

To Calculate $\overline{X}$ add all the observations (scores) and divide by the number of observations:

$$\overline{X} = \frac{81+76+51+65+70+85+91}{8}$$

$$= \frac{1434}{8}$$

$$= 72.25$$

RULE:

$$t = \frac{(\overline{X} - \mu)}{\frac{S_X}{\sqrt{n-1}}}$$

(2) When you have a *two-tailed rejection region,* fail to reject the null hypothesis if the |*calculated value*| > critical value.

Calculated t = 0.56

Critical t = 2.365

Comparing the value of critical t and the calculated value in a two-tailed distribution:

0.56 < 2.365 → Cal t < critical value → reject H_0

CONCLUSION:

The mean of the sample (72.25) is significantly different from the hypothesized mean (75). The sample results do not match the hypothesized mean.

THE RELATIONSHIP BETWEEN HYPOTHESIS & SIGNIFICANCE:

It is important to understand that the significance of the results is dependent entirely on the hypothesis.

EXAMPLE 2:

Hour	gals per min
1	81
2	76
3	51
4	65
5	56
6	73
7	85
8	91

Figure 59a: Unit 4C-3, Example 2, data.

Suppose that for this example, we use the same data from **Example 2.** The differences: The data is from hourly tests of a hydrothermal flow. The hypothesis is that the flow will average *at least* 75 gallons per minute that is, the flow will be greater than 75 gals per minute. Over a period of 8 hours, the results:

Hypotheses:

$$H_0: \overline{X} \geq 75$$

$$H_a: \overline{X} < 75$$

This is a left-tailed, one-tailed distribution, therefore, if |calculated value| < critical value, fail to reject **H₀**.

The data has not changed, but the hypothesis has. The type of distribution is different which changes the rule.

STEP 1:

To Calculate $\overline{X}$:

Add all the observations (scores) and divide by the number of observations

$$\overline{X} = \frac{81+76+51+65+70+85+91}{8}$$

$$= \frac{1434}{8}$$

$$= 72.25$$

STEP 2:

Subtract the hypothesized mean (μ) from the actual mean (**0**).

72.25 - 75 = -2.75

STEP 3:

Calculate the square root of the total number of observations.

√8 = 2.83 (to 2 decimal places)

STEP 4:

Multiply the results of steps 2 and 3.

-2.25 * 2.83 = -6.37 (to 2 decimal places)

STEP 5:

Calculate S_x the standard deviation

- The mean $\overline{X}$ = 72.25
- Subtract $\overline{X}$ from each observation (col 3).
- Square each $(X-\overline{X})$ col 4).
- Find $\Sigma(X-\overline{X})^2$ (sum each observation).
- Divide $\Sigma(X-\overline{X})^2$ by (n-1)

  1364.98 ÷ 7 = 195.00

- Find the square root of $\Sigma(X-\overline{X})^2$ ÷ (n-1)

Hour	gals/min	(X-X)	(X-X)²
1	81	(81-72.25)= 8.25	68.06
2	76	(76-72.25)= 3.75	14.06
3	51	(51-72.25)=-21.25	451.56
4	65	(65-72.25)= -7.25	52.56
5	56	(56-72.25)=-16.25	264.06
6	73	(73-72.25)= 0.75	0.56
7	85	(85-72.25)= 12.75	162.56
8	91	(91-72.25)= 18.75	351.56
			Σ(X-X̄)² = 1364.98

S_x = √[195.00] = 13.96

Figure 54b: Unit 4C-3, Example 2, Step 5.

STEP 6:

Calculate the t-value by dividing the result of step 4 by the standard deviation.

Step 1: $\overline{X}$ = 72.25
Step 3: √n = 2.83
Step 5: S_x = 13.96
μ = hypothesized mean = 75
df = 8-1 = 7

$$t = \frac{(\overline{X} - \mu) * \sqrt{n}}{S_x}$$

t = {(72.25-75)*2.83} ÷ 13.96

= {-2.75*2.83} ÷ 13.96

= {7.78} ÷ 13.96

= -0.56

Calculated |t| = |-0.56|
= 0.56

Hypotheses:

$H_0: \overline{X} \geq 75$

$H_a: \overline{X} < 75$

The null and alternative hypotheses give us a one-tailed distribution with 7 degrees of freedom.

t Table

one-tail	0.50	0.25	0.20	0.15	0.10	0.05	0.025	0.01
two-tails	1.00	0.50	0.40	0.30	0.20	0.10	0.05	0.02
df 1	0.000	1.000	1.376	1.963	3.078	6.314	12.71	31.82
2	0.000	0.816	1.061	1.386	1.886	2.920	4.303	6.965
3	0.000	0.765	0.978	1.250	1.638	2.353	3.182	4.541
4	0.000	0.741	0.941	1.190	1.533	2.132	2.776	3.747
5	0.000	0.727	0.920	1.156	1.476	2.015	2.571	3.365
6	0.000	0.718	0.906	1.134	1.440	1.943	2.447	3.143
7	0.000	0.711	0.896	1.119	1.415	1.895	2.365	2.998
8	0.000	0.706	0.889	1.108	1.397	1.860	2.306	2.896
9	0.000	0.703	0.883	1.100	1.383	1.833	2.262	2.821
10	0.000	0.700	0.879	1.093	1.372	1.812	2.228	2.764
11	0.000	0.697	0.876	1.088	1.363	1.796	2.201	2.718

Figure 55c: Unit 4C-3, example 3, two tails, df = 7

Rule: When you have a one-tailed rejection region, fail to reject the null hypothesis if the |calculated value| < critical value.

Calculated t = |0.56| and from the table, critical $t_{(df)(\alpha)}$ = critical $t_{(7)(0.05)}$ = 1.895.

0.56 < 1.895

So fail to reject the null hypothesis.

CONCLUSION:

The mean of the sample set is not significantly different from the hypothesized mean; that is, the average flow of the sample data is equivalent to 75 gallons per hour.

COMPARISON OF EXAMPLE 1 & EXAMPLE 2:

The data for each example is the same although the type of data is different. In Example 2, the data was student scores; in Example 3 the data was gallons per minute. The purpose of this exercise was to show that **the hypotheses** are what cause the difference in results. Indeed, the nature of the data involved required different hypotheses which change a two-tail test to a one tail test; even though the data was the same.

The hypotheses:

Example 1: $H_0: \overline{X} = \mu$

$H_a: \overline{X} \neq \mu$

The mean has a single value and the distribution has two tails.

Rule 2 applies: When you have a *two-tailed rejection region,* fail to reject the null hypothesis if the |*calculated value*| > critical value.

Calculated t = 0.56; Critical t = 2.262

0.56 < 2.262, **reject H₀**

Example 2: $H_0: \overline{X} \geq 75$

$H_a: \overline{X} < 75$

The mean has a range of values and the distribution has a single tail.

Rule 1 applies: *When* you have a *one-tailed rejection region,* fail to reject the null hypothesis if the *calculated value < critical value.*

Calculated t = 0.56; Critical t = 2.262

0.56 < 2.262, **0 fail to reject H₀**

Analysis: The different hypotheses require the use of a different rule, resulting in the difference in whether H₀ is rejected or not.

EXAMPLE 3:

A final example of the ***group to population t-test***: At a breakfast cereal plant, at one automatic filling machine the desired weight for each carton was 16 oz. A random sample of 15 cartons was collected and the contents weighed. The mean weight of the cartons was 15.8 oz. The standard deviation of the sample (**S**) was measured at 0.15 oz. Does the sample differ significantly from the desired weight?

INFORMATION SUMMARY:

n = 15

0 = 15.8 oz.

S_x = 0.15 oz.

α = 0.05

$H_0: \mu$ = 16

$H_a: \mu$ ≠ 16

Rule: Two-tailed distribution: If calculated t > critical t, reject **H₀**.

SOLUTION:

$$t = \frac{(\bar{X} - \mu) * \sqrt{n}}{S_X}$$

$$t = \frac{(15.8 - 16) * \sqrt{9}}{0.15}$$

$$= \frac{-0.2 * 3}{0.15}$$

$$= \frac{-0.6}{0.15}$$

$$= -4$$

t Table

one-tail	0.50	0.25	0.20	0.15	0.10	0.05	0.025
two-tails	1.00	0.50	0.40	0.30	0.20	0.10	0.05
df							
1	0.000	1.000	1.376	1.963	3.078	6.314	12.71
2	0.000	0.816	1.061	1.386	1.886	2.920	4.303
3	0.000	0.765	0.978	1.250	1.638	2.353	3.182
4	0.000	0.741	0.941	1.190	1.533	2.132	2.776
5	0.000	0.727	0.920	1.156	1.476	2.015	2.571
6	0.000	0.718	0.906	1.134	1.440	1.943	2.447
7	0.000	0.711	0.896	1.119	1.415	1.895	2.365
8	0.000	0.706	0.889	1.108	1.397	1.860	2.306
9	0.000	0.703	0.883	1.100	1.383	1.833	2.262
10	0.000	0.700	0.879	1.093	1.372	1.812	2.228
11	0.000	0.697	0.876	1.088	1.363	1.796	2.201
12	0.000	0.695	0.873	1.083	1.356	1.782	2.179
13	0.000	0.694	0.870	1.079	1.350	1.771	2.160
14	0.000	0.692	0.868	1.076	1.345	1.761	2.145
15	0.000	0.691	0.866	1.074	1.341	1.753	2.131

Figure 56: two-tails, df=14.

Since |-4| = 4; Critical t at 95% (0.05) with n-1 degrees of freedom (15-1 = 9):

Critical $t_{(15)(0.05)}$ = 2.145

Calculated t = 4

Calculated |t| > Critical t

RULE:

When you have a two-tailed rejection region, fail to reject the null hypothesis if the |calculated value| > critical value.

Since 4 > 2.262, fail to reject **H₀**

CONCLUSION:

Since calculated t is greater than critical t, there is no significant difference.

The automatic filling machine does not need to be adjusted.

UNIT 4C-4: THE TWO GROUP DEPENDENT t-test.

PURPOSE:

There are two t-tests that compare observations from two groups. One is where the groups being studied are dependent; the other is where the two groups are independent. Conducting a two group t-test means using the values obtained through actual observations then calculating a t-value (no hypothesized mean). This t-value is used to draw a conclusion concerning whether the hypothesis is true or not. This lesson considers the two group dependent t-test.

Dependent samples occur when you collect data from the same group of subjects. You may compare data collected at two different times, such as the results of the mid-term exam and the final. Or you may compare the data for the same set of students in two different subjects such as Math and English. An example of this type of test is the Graduate Record Exam (GRE), where the subjects are tested in both Math and English and the results compared. What is important is that the same group of participants is involved in the two sets of data.

OBJECTIVES:

- Compare two sets of dependent samples using the t-test.

- Calculate the variance and standard deviations.

- Use the calculated value and the critical value to make a conclusion about the data.

TWO GROUP DEPENDENT (PAIRED DATA):

We may wish to obtain observations for the same subjects at two different times to determine if the results are different. For example, we may hypothesize that the salary of a group of graduates, before and after obtaining their Master's degree, remains the same. Here the data or observations are obtained for the same set of people at two different times — their annual earnings before and again after obtaining their degree. Such data is called paired data. A t-test may be used to compare the two groups of observations and to test the null hypothesis that they are equal, in other words, that obtaining the degree does not make any difference in salary.

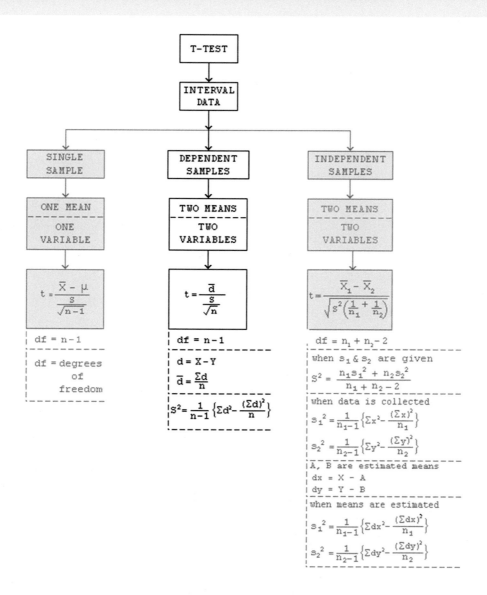

Figure 57: Emphasizing the t-test for dependent samples.

THE t-VALUE:

The t-test when the two groups are dependent considers two groups with the same members but two different sets of observations. As with all such tests, a null hypothesis and an alternate hypothesis must be declared.

Since actual observations are required, there are two variables, X and Y. Also, since the same subjects are being observed, n represents the number of members of both groups.

The formula for the variance in the flow chart is:

Where d = X - Y, the difference between each participant's two observations (paired). The mean of d is the sum of these differences divided by n.

S is the standard deviation based on the differences between the pairs of observations.

EXAMPLE 1:

A test was given to 5 students at the beginning of the term (bt) and again at the end of the term (at).

$$H_0: X_{bt} = X_{at}$$

$$H_a: X_{bt} \neq X_{at}$$

This is a two-tailed distribution because X_{bt} could be **either** less than **or** greater than X_{at}. Fail to reject the null hypothesis if the |*calculated value*| > critical value.

Test for significance at $\alpha = 0.05$ and at $\alpha = 0.01$.

Determine whether the difference between the test results (bt-at) is significant.

Students	bt	at
1	90	95
2	85	88
3	83	85
4	79	86
5	95	99

Figure 58a: Unit 4C-4, Example 1, data.

Students	bt	at	d = bt-at	d^2
1	90	95	5	25
2	85	88	3	9
3	83	85	2	4
4	79	86	7	49
5	95	99	4	16
			$\Sigma d=21$	$\Sigma d^2=103$

Figure 58b: Unit 4C-4, Example 1, calculate d and d^2.

$$t = \frac{\bar{d}}{S \div \sqrt{n-1}}$$

$$S = \sqrt{\frac{1}{n-1}\left\{\sum d^2 - \frac{(\sum d)^2}{n}\right\}}$$

$$= \sqrt{\frac{1}{4}\left[103 - \frac{(21)^2}{5}\right]}$$

$$= \sqrt{\frac{1}{4}\left[103 - \frac{441}{5}\right]}$$

$$= \sqrt{\frac{1}{4}\left[103 - 88.2\right]}$$

$$= \sqrt{\frac{14.8}{4}}$$

$$= \sqrt{3.7}$$

$$= 1.9235$$

→ S = √3.7 = 1.92 to 2 decimal places

The mean of d:

$$\bar{d} = \frac{\sum d}{n}$$

= 21/5

= 4.2

The t-value:

$$t = \frac{\bar{d}}{S \div \sqrt{n-1}}$$

= 4.2 ÷ [1.92 ÷ √4]

= 4.2 ÷ [1.92 ÷ 2]

= 4.2 ÷ 0.96

= 4.375

t Table

one-tail	0.50	0.25	0.20	0.15	0.10	0.05	0.025	0.01	0.005
two-tails	1.00	0.50	0.40	0.30	0.20	0.10	0.05	0.02	0.01
df 1	0.000	1.000	1.376	1.963	3.078	6.314	12.71	31.82	63.66
2	0.000	0.816	1.061	1.386	1.886	2.920	4.303	6.965	9.925
3	0.000	0.765	0.978	1.250	1.638	2.353	3.182	4.541	5.841
4	0.000	0.741	0.941	1.190	1.533	2.132	2.776	3.747	4.604
5	0.000	0.727	0.920	1.156	1.476	2.015	2.571	3.365	4.032
6	0.000	0.718	0.906	1.134	1.440	1.943	2.447	3.143	3.707
7	0.000	0.711	0.896	1.119	1.415	1.895	2.365	2.998	3.499
8	0.000	0.706	0.889	1.108	1.397	1.860	2.306	2.896	3.355
9	0.000	0.703	0.883	1.100	1.383	1.833	2.262	2.821	3.250
10	0.000	0.700	0.879	1.093	1.372	1.812	2.228	2.764	3.169
11	0.000	0.697	0.876	1.088	1.363	1.796	2.201	2.718	3.106
12	0.000	0.695	0.873	1.083	1.356	1.782	2.179	2.681	3.055
13	0.000	0.694	0.870	1.079	1.350	1.771	2.160	2.650	3.012
14	0.000	0.692	0.868	1.076	1.345	1.761	2.145	2.624	2.977
15	0.000	0.691	0.866	1.074	1.341	1.753	2.131	2.602	2.947

Figure 58c: example 1, critical value for 4 df, at α = 0.05 and at α = 0.01.

Significance at α = 0.05

Calculated value = 4.375

Critical value at α = 0.05 → 2.132

Rule: Two-tailed test: fail to reject H_0 if calculated value > critical value.

Calculated t > critical t → fail to reject H0

Conclusion: The end of term exam results are not significantly different from the beginning of the term results at **α = 0.05**.

Significance at α = 0.01

Calculated value = 4.375

Critical value at α = 0.01 → 4.604

Rule: Two-tailed test: fail to reject H_0 if calculated value > critical value.

Calculated t < critical t → reject H_0

Conclusion: The end of term exam results are significantly different from the beginning of the term results **α = 0.01**.

When working with a dependent two-group t test, always check BOTH significance levels because the greater significance may enable you to reject H_0.

EXAMPLE 2:

Ten students were tested twice during a seminar on the same topic. Were the scores for the second test significantly different than their scores on the first test?

We start with the alternative hypothesis because the presenter before the seminar wants the second test to show improvement. Therefore the inequality chosen is < (not ≤ as the second test should show improvements):

H_a: **Exam 1 Scores** < **Exam 2 Scores**

The null hypothesis uses the opposite equality. That is, if there is no improvement, the scores for Exam 1 should be at least as good as, or possibly better than Exam 2 scores.

H_0: **Exam 1 Scores** ≥ **Exam 2 Scores**

Given information:

$\sum d = 27$

$\sum d/n = 27/10 = 2.7$ (the mean)

$\sum d^2 = 729$

$\sqrt{n} = 3.16$

$\alpha = 0.05$

Students	Exam 1	Exam 2
1	82	83
2	86	88
3	79	82
4	91	95
5	89	90
6	75	77
7	84	89
8	73	78
9	84	89
10	77	77

Figure 59a: Unit 4C-4, Example 2, data.

Calculate the variance:

$$S^2 = \frac{1}{n-1}\left\{\sum d^2 - \frac{\left(\sum d\right)^2}{n}\right\}$$

$S^2 \quad = 1 \div 9 * \{101 - \{(54)^2 \div 10\}$
$\quad\quad = 1 \div 9 * \{101 - 2916/10\}$
$\quad\quad = 0.11 * \{101 - (291.6\}$
$\quad\quad = 0.11 * \{89.6\}$
$\quad\quad = 9.86$ to 2 decimal places

$S = \sqrt{9.86} = 3.14$

$\sum d = 54$

$\sum d/n = 54/10 = 5.4$ (the mean)

$\sqrt{(n)} = \sqrt{10} = 3.16$

$\alpha = 0.05$

$$t = \frac{\bar{d}}{S \div \sqrt{n-1}}$$

Students	Exam 1 X	Exam 2 Y	d Y – X	d² (Y – X)²
1	82	83	1	1
2	86	88	2	4
3	79	82	3	9
4	91	95	4	16
5	89	90	1	1
6	75	77	2	4
7	84	89	4	16
8	73	78	5	25
9	84	89	5	25
10	77	77	0	0
n=10			$\sum d = 54$	$\sum d^2 = 101$

Figure 59b: Unit 4C-4, Example 2, calculation.

$= 2.7 \div (3.14 \div 3.16) = 2.7 \div 0.99 = 2.72$

Partial t-table

t Table								
one-tail		0.50	0.25	0.20	0.15	0.10	0.05	0.025
two-tails		1.00	0.50	0.40	0.30	0.20	0.10	0.05
df	1	0.000	1.000	1.376	1.963	3.078	6.314	12.71
	2	0.000	0.816	1.061	1.386	1.886	2.920	4.303
	3	0.000	0.765	0.978	1.250	1.638	2.353	3.182
	4	0.000	0.741	0.941	1.190	1.533	2.132	2.776
	5	0.000	0.727	0.920	1.156	1.476	2.015	2.571
	6	0.000	0.718	0.906	1.134	1.440	1.943	2.447
	7	0.000	0.711	0.896	1.119	1.415	1.895	2.365
	8	0.000	0.706	0.889	1.108	1.397	1.860	2.306
	9	0.000	0.703	0.883	1.100	1.383	1.833	2.262
	10	0.000	0.700	0.879	1.093	1.372	1.812	2.228
	11	0.000	0.697	0.876	1.088	1.363	1.796	2.201
	12	0.000	0.695	0.873	1.083	1.356	1.782	2.179
	13	0.000	0.694	0.870	1.079	1.350	1.771	2.160

Figure 59c: example 2, critical value for df = 9, where $\alpha = 0.05$ and $\alpha = 0.01$.

FIRST α:
df = 9, α = 0.05
Critical t = $t_{(df)(α)}$ = 2.262

SECOND α:
df = 9, α = 0.01
Critical t = $t_{(df)(α)}$ = 3.250

Rule: When you have a two-tailed rejection region, fail to reject the null hypothesis if the |calculated value| > critical value.

FIRST α (0.05):

Critical t = 2.262

Calculated t = 2.72

Calculated t > critical t

For a two-tail distribution, calculated t > critical t → fail to reject H_0

→ There is NO significant difference between the first test scores and the second test scores at a significance level of 0.05.

FIRST α: CONCLUSION:

The seminar did not significantly improve the students' scores.

SECOND α (0.01):

Critical t = 3.250

Calculated t = 2.72

|Calculated t| < critical t

For a two-tail distribution, |calculated t| < critical t **0** reject H_0

SECOND α: CONCLUSION:

The seminar improved the students' scores at a significance level of 0.01.

Why the difference? Remember the relationship for the two-tailed t-test is an inverse relationship so, as the level of significance decreases (0.05 down to 0.01) the more likely that the null hypothesis will be rejected.

THE IMPORTANCE OF THE LEVEL OF SIGNIFICANCE:

Suppose the calculated value of t had been 3.7 instead of 4.28 with the same degrees of freedom (4). Then for a one-tail distribution:

t Table											
one-tail	0.50	0.25	0.20	0.15	0.10	0.05	0.025	0.01	0.005	0.001	0.0005
two-tails	1.00	0.50	0.40	0.30	0.20	0.10	0.05	0.02	0.01	0.002	0.001
df 1	0.000	1.000	1.376	1.963	3.078	6.314	12.71	31.82	63.66	318.31	636.62
2	0.000	0.816	1.061	1.386	1.886	2.920	4.303	6.965	9.925	22.327	31.599
3	0.000	0.765	0.978	1.250	1.638	2.353	3.182	4.541	5.841	10.215	12.924
4	0.000	0.741	0.941	1.190	1.533	2.132	2.776	3.747	4.604	7.173	8.610
5	0.000	0.727	0.920	1.156	1.476	2.015	2.571	3.365	4.032	5.893	6.869
6	0.000	0.718	0.906	1.134	1.440	1.943	2.447	3.143	3.707	5.208	5.959
7	0.000	0.711	0.896	1.119	1.415	1.895	2.365	2.998	3.499	4.785	5.408
8	0.000	0.706	0.889	1.108	1.397	1.860	2.306	2.896	3.355	4.501	5.041
9	0.000	0.703	0.883	1.100	1.383	1.833	2.262	2.821	3.250	4.297	4.781
10	0.000	0.700	0.879	1.093	1.372	1.812	2.228	2.764	3.169	4.144	4.587
11	0.000	0.697	0.876	1.088	1.363	1.796	2.201	2.718	3.106	4.025	4.437
12	0.000	0.695	0.873	1.083	1.356	1.782	2.179	2.681	3.055	3.930	4.318
13	0.000	0.694	0.870	1.079	1.350	1.771	2.160	2.650	3.012	3.852	4.221
14	0.000	0.692	0.868	1.076	1.345	1.761	2.145	2.624	2.977	3.787	4.140
15	0.000	0.691	0.866	1.074	1.341	1.753	2.131	2.602	2.947	3.733	4.073
16	0.000	0.690	0.865	1.071	1.337	1.746	2.120	2.583	2.921	3.686	4.015
				1.069	1.333	1.740	2.110	2.567	2.898	3.646	3.965
							2.101	2.552	2.878	3.610	3.922

Figure 63c -Unit 4C-4, e.g. 2, 2-tailed. df=4, at significance levels of 0.05 (red) and 0.01 (green).

The table in Figure 63c shows the two tail critical values for α = 0.05 (red markings) and α = 0.01 (blue markings).

Critical value α_{red} at 0.05 → 2.132

Calculated Value = 3.7

3.7 > 2.132 → fail to reject H_0 [at .0.05)

Critical value α_{blue} at 0.01 → 3.747

Calculated Value = 3.7

3.7 < 3.747 → reject H_0 [at 0.01]

You might think that the difference of 0.047 is not a significant amount but, never-the-less, 3.7 is less than 3.747! Therefore we would reject H_0. On the other hand if the calculated value of t had been 3.8, or more, then the null hypothesis would fail to be rejected at both levels of significance.

Another factor in finding the critical value is the number of degrees of freedom. Had there been 5 degrees of freedom in this problem instead of 4, both critical values (2.016 and 3.365) would have shown that you could reject the null hypothesis at either level of significance.

The third factor as far as the t-test is concerned is whether the curve shows one tail or two. Notice that the same level of significance generally shows the two-tail information (red arrows) is one column to the right of the one-tail information (green arrows).

In any study you complete, being able to reject the null hypothesis at 0.01 will make the results of your study much more significant than at 0.05. But 0.05 is an acceptable significance level as it means that there is a 95% chance that your sample fits the population from which your sample is drawn so you can generalize the findings to the population. Of course, that also means that there is a 5% chance that your sample doesn't fit the population. But 0.05 is an acceptable level of significance.

UNIT 4C-5: THE TWO GROUP INDEPENDENT t-TEST.

PURPOSE:

We have discussed how to analyze a single variable and mean as well as how to analyze two mean when you have two dependent groups. To calculate the **t-value** when there was only one variable, you compared the mean for that variable with a population mean or a hypothesized mean. The difference between the one-variable t test and the other two t-tests is that each of the two-group t-tests have two observed means. The dependent groups have the same members being observed at two different times or for two different topics. The independent group has two unrelated sets of members and the topics studied for each group is the same.

This lesson discusses the hypothesis test for the difference between two means where the two samples are independent, that is, the participants in one group are totally unrelated to the participants in the second group. The test procedure is called the two-sample independent t-test and there are certain conditions.

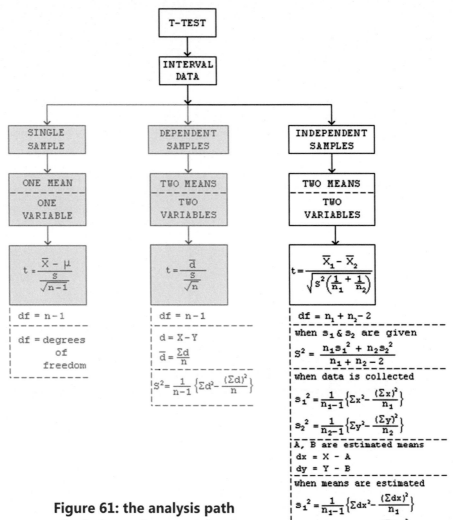

Figure 61: the analysis path for independent samples.

OBJECTIVES:

- Calculate the t-value for a two-group set of unpaired and independent observations.

- Perform the independent two-sample t-test on your own data. You will need two different sets of participants using the same test for each group.

INTRODUCTION:

This lesson discusses the hypothesis test for the difference between two means when two samples are independent. The meaning of "independent" in this circumstance is that the samples are not related. For instance: suppose you are studying the effect of gaining a master's degree on the salary. If you compare the salary of a group of graduates before and after the master's degree, the groups are dependent because you gather the information from the same group of people. If, instead, you compare the salaries of graduate students from two different universities following the master's degree, the two sample are independent. In other words there are no subjects attending the first university in the sample from the second university or vice versa. Independence means that no subject from the first group is in the second group, nor are there subjects from the second group in the first.

There are certain conditions in order to make sure the two groups in your study are independent.

- Simple random sampling must be used to find the members of the two samples.

- The two samples must be independent, that is, there is no overlap or relationship between members of the two samples.

- The populations from which each sample is drawn are as close to normal as possible.

- Each sample size should be either:

    ○ 15 or less with unimodal (single mode distribution) and without outliers.

    ○ If there are 16 to 40 members in each sample, it may be slightly skewed but should still be unimodal and have no outliers.

    ○ If the sample size if greater than 40, there must be no outliers.

    ○ [Reminder: Outliers are data points that lie outside the normal range of data.]

TWO INDEPENDENT SAMPLES (TWO UNPAIRED GROUPS):

There is just the one formula used for S^2 when you are working with independent groups as shown in the flow chart. However, there are three situations that may be involved.

The first is where you have two independent groups and are given the mean and variance and you are asked to determine if there is a difference between the two groups.

The second is where you have collected data so the variance for each group is found by using the given data.

The third is where you have data for each group and use estimated means to determine the variances. The advantage is that the estimated means are whole numbers.

Note: because the two groups are independent, there may be a different number of participants in each group. The general formula for the t value when dealing with an independent two-group sample is:

$$t = \frac{\overline{X} - \overline{Y}}{\sqrt{S_T^2 \left[\frac{1}{n_1} + \frac{1}{n_2} \right]}}$$

Where $\overline{X}_1$ and $\overline{X}_2$ are the means for the respective groups.

S_T is the total standard deviation.

n_1 and n_2 are the numbers of observations per group.

$df = n_1 + n_2 - 2$

s_1 = Standard Deviation for Group 1

s_2 = Standard Deviation for Group 2

First let us look at the situation where you are given the means and the standard deviations for the two groups. Note that a lower case **s** is used for the standard deviation for each of the two variables. Note also that there are three conditions required for performing this t-test.

1. $n \geq 30$

2. The populations are normally distributed.

3. Population variance (σ^2) is unknown.

EXAMPLE OF FIRST SITUATION – MEANS & STANDARD DEVIATION GIVEN:

There are 38 voting districts in Town A. The mean (μ_1) for the number of voters at each of the districts in an election is 15; and the standard deviation is 2. Town B has 30 voting districts with a mean (μ_2) of 14 and a standard deviation of 2.5. Is there a significant difference between the voters in the two towns?

CALCULATION:

State the hypotheses and the level of significance:

$H_0: \mu_1 = \mu_2$

$H_a: \mu_1 \neq \mu_2$

$\alpha = 0.05/0.01/0.10$ (any of these significance values is acceptable but we will use 0.05)

Given data:

Town A	Town B
$n_1 = 38$	$n_2 = 30$
$\bar{x}_1 = 15$	$\bar{x}_2 = 14$
$S_1 = 2$	$S_2 = 2.5$

$df = n_1 + n_2 - 2$

$df = 38 + 30 - 2 = 66$

$$S_T^2 = \frac{n_1 S_1^2 + n_2 S_2^2}{n_1 + n_2 - 2}$$

$= \{38(2)^2 + 30(2.5)^2\} \div \{38 + 30 - 2\}$
$= \{38(4) + 30(6.25)\} \div \{66\}$
$= \{152 + 187.5\} \div \{66\}$
$= \{339.5\} \div \{66\}$
$= 5.14$

$$t = \frac{\bar{X} - \bar{Y}}{\sqrt{S_T^2 \left[\frac{1}{n_1} + \frac{1}{n_2} \right]}}$$

$= (15 - 14\} \div (5.14) \sqrt{(0.026 + 0.033)}$
$= 1 \div \sqrt{(0.059)}$
$= 1 \div 0.24$
$= 4.166667$
$= 4.17$ To 2 Decimal Places.

To determine if this value for t has significance, it is necessary to use the t-table. It is also necessary to determine if the distribution has one tail or two. To do that, look at the null and alternative hypotheses:

$H_0: \mu_1 = \mu_2$

$H_a: \mu_1 \neq \mu_2$

H_a indicates that the value of u_1 and u_2 lie on either side of the value for H_0. This is a two tail test. By stating the alternate hypothesis this way we are saying in effect that μ_1 and μ_2 are outside of the range defined by the null hypothesis.

Note that if the null hypothesis contained an inequality, such as $\mu_1 \leq \mu_2$ or $\mu_1 \geq \mu_2$ the t-test has one tail, the tail contains all values greater than μ_1 with $\leq$; or all values less than μ_1 with $\geq$.

The direction of the tail is indicated by whether ≤ or ≥ is the inequality used – the tail is on the "open side" of the inequality. In the case of ≤ the tail is to the right; in the case of ≥ the tail is to the left.

PROBLEMS WITH THE t-TABLE CHART:

There is a problem with the table we have been using. Below is the bottom half of this t-table:

Look at the column showing the degrees of freedom. The values to 30 are inclusive, but from 30, only the major values of 30, 40, 60, 80, and 100 are present (see Figure 63c).

One can "guess" that since the value we need (66) is close to half way between 60 and 80, the critical value we need would be between (two tails) 1.990 and 2.000. *But we should not estimate the critical value; we need this value to be exact.* So, for df values greater than 30, we have to look for another way of determining the critical value

16	0.000	0.690	0.865	1.071	1.337	1.746	2.120	2.583	2.921	3.686	4.015
17	0.000	0.689	0.863	1.069	1.333	1.740	2.110	2.567	2.898	3.646	3.965
18	0.000	0.688	0.862	1.067	1.330	1.734	2.101	2.552	2.878	3.610	3.922
19	0.000	0.688	0.861	1.066	1.328	1.729	2.098	2.539	2.861	3.579	3.883
20	0.000	0.687	0.860	1.064	1.325	1.725	2.086	2.528	2.845	3.552	3.850
21	0.000	0.686	0.859	1.063	1.323	1.721	2.080	2.518	2.831	3.527	3.819
22	0.000	0.686	0.858	1.061	1.321	1.717	2.074	2.508	2.819	3.505	3.792
23	0.000	0.685	0.858	1.060	1.319	1.714	2.069	2.500	2.807	3.485	3.768
24	0.000	0.685	0.857	1.059	1.318	1.711	2.064	2.492	2.797	3.467	3.745
25	0.000	0.684	0.856	1.058	1.316	1.708	2.060	2.485	2.787	3.450	3.725
26	0.000	0.684	0.856	1.053	1.315	1.706	2.056	2.479	2.779	3.435	3.707
27	0.000	0.684	0.855	1.057	1.314	1.703	2.052	2.473	2.771	3.421	3.690
28	0.000	0.683	0.855	1.056	1.313	1.701	2.043	2.467	2.763	3.408	3.674
29	0.000	0.683	0.854	1.055	1.311	1.699	2.045	2.462	2.756	3.396	3.659
30	0.000	0.683	0.854	1.055	1.310	1.697	2.042	2.457	2.750	3.385	3.646
40	0.000	0.681	0.851	1.050	1.303	1.684	2.021	2.423	2.704	3.307	3.551
60	0.000	0.679	0.848	1.045	1.296	1.671	2.000	2.390	2.660	3.232	3.460
80	0.000	0.678	0.846	1.043	1.292	1.664	1.990	2.374	2.639	3.195	3.416
100	0.000	0.677	0.845	1.042	1.290	1.660	1.984	2.364	2.626	3.174	3.390
1000	0.000	0.675	0.842	1.037	1.282	1.646	1.962	2.330	2.531	3.093	3.300
	0%	50%	60%	70%	80%	90%	95%	98%	99%	99.8%	99.9%
					Confidence Level						

Figure 63: Bottom half of t-table showing the df limitation.

INTERNET HELP:

There are several on-line sites that assist in giving you the process for determining the critical t-value.

http://www.itl.nist.gov/div898/handbook/eda/section3/eda3672.htm

Select 1.3.6.7. Tables for Probability Distributions which takes you to the bottom of the page.

Click on "NEXT" then go to the bottom of this page and click on "NEXT" again.

1.3.6.7.2. Critical Values of the Student's *t* Distribution

Move down the page until you see the table. This site shows the full table but only for one-tail values. Check the df of 66; the second column shows the significance of 0.05 for a one-tailed distribution. So, you can guess that the critical value for a two-tail distribution with 66 degrees of freedom is about 1.997 (remember two-tail values are one column to the right). **BUT GUESSING IS NOT ACCEPTABLE.**

There are other sites that give you all the information you need.

THE HOMEWORK.COM SITE:

This is probably one of the most useful of the t-table sites as you can use it to calculate the critical t value for left- and right-tailed data as well as middle and two-tailed data. It also shows the areas under the curve for each. Note: you have already looked at this site earlier so this is a review of what you need to know.

http://www.tutor-homework.com/statistics_tables/statistics_tables.html

Note: At this site, the first table calculator is for the z distribution; the second is for the t distribution and the third is for the Chi-Square distribution.

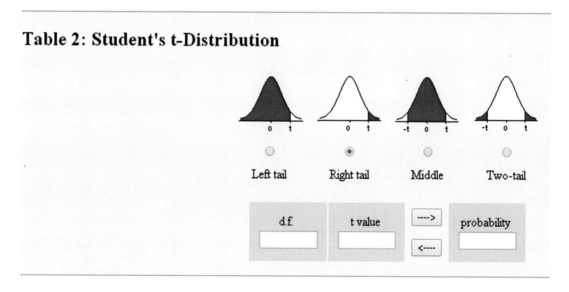

Figure 64: Online calculator for calculating critical-t or the probability.

For the t-table calculator, you may select left- tail, right-tail, the middle or the two-tail (red areas in each graphic of a curve are areas of rejection of the null hypothesis.). *When using this calculator, you are either looking for the probability P, or the t-value. Your hypothesis determines which tail is to be used.*

Null hypothesis analyzed:

$H_0: \mu_1 \geq t;$ **one tail (left tailed).**

$H_0: \mu_1 \leq t;$ **one tail (right tailed).**

$H_0: -\mu_1 \leq H_0 \leq \mu_2;$ **middle.**

$H_0: \mu_1 = \mu_2;$ **Two tail.**

THE T-DISTRIBUTION CALCULATOR:

Results from this calculator are determined by whether you select "left tail," "right tail," "middle" or "two-tail." The default at the site is for "right tail." One of the two arrows for the calculation is selected based on the direction of the calculation: If you wish to find the probability of the t-value existing, you enter the t-value and select the top arrow (pointing right) to find the probability. If you have a specific value for the probability, such as 0.05, you enter that value and press the lower arrow (pointing left). This will give you the critical value for t. Compare it with the calculated value then use the three rules on page 77, to determine if you can reject the null hypothesis.

When using this distribution calculator, the first thing you need to do is determine which region you need to select. You must also use the degrees of freedom. In the first situation, df = 38 + 30 -2 or 66 and the distribution is two-tailed. Your significance level (probability) is 0.05 and your set up for the problem will look like Figure 68a.

Examples of using the homework site's t-calculator.

Select the Two-tail option.

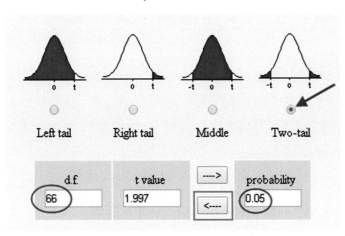

Figure 65a: 2-tail example, 66 df, probability = 0.05.

Write in 66 degrees of freedom.

Write in 0.05 as the probability (level of significance).

Press the left-pointing arrow and the calculator shows a critical t-value of 1.997.

If you wish to know the probability of the calculated t-value, 4.17, occurring, set up the problem this way by inserting the calculated t of 4.17, press the upper arrow and the probability is 0.0001, that is, a probability of 1/10000% a miniscule amount. Since you need a probability of at least 90% (0.900), you fail to reject H_0.

**Figure 65b: 2-tail example,
66 df t-value of 4.17.**

COMPARING RESULTS SELECTING DIFFERENT TAILS:

If you are not already at the tutor-homework site, use the following to get to the web site and click on the second calculator: Student's t-distribution.

http://www.tutor-homework.com/statistics_tables/statistics_tables.html

The right-tail is the default selection. First insert the data for items 1, 2 & 3 in the t-distribution calculator. Then press the button labeled 4 to get the result.

1. Select the right tail option.

2. Write in 18 df.

3. Insert the probability of 0.05.

4. Press the left arrow.

RESULT: 1.734.

Table 2: Student's t-Distribution

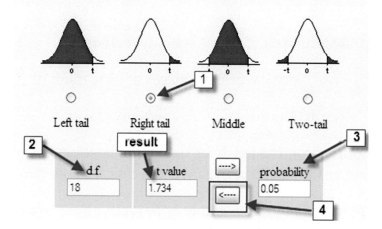

**Figure 65c: independent test t-value,
df =18, probability = 0.05.**

NOW COMPARE THE EFFECT OF MAKING THE FOLLOWING CHANGES:

1. Select the left tail dot. This is for the H_0: $\mu_x \leq \mu_y$

   Result: The same value results as from the right tail choice except that it is now a negative value.

2. Select the "middle" button. This is for the hypothesis H_0: $-\mu_{x1} \leq \mu_y \leq +\mu_{x2}$ (μ_y lies between the low and high values of μ_x).

   Result: Critical t is now 0.06359.

3. Select the two-tail dot. The hypothesis is H_0: $-\mu_x = \mu_Y$

   Result: Critical t is now 2.101.

4. Set the selection to the right tail (arrow 1). The hypothesis is H_0: $\mu_x \geq \mu_y$. Make the probability 0.01 (arrow 3) that is 99%.

   Result: Critical t is now 2.552. The critical value always increases when the probability is greater (99% is greater than 95%).

5. With the right tail selected and the hypothesis being H_0: $\mu_x \geq \mu_y$

   Result: Make the probability 0.10 (90%). Critical t is now: 1.3. The critical value always decreases when the probability is less (90% is less than 95% or 99%).

6. Set the probability back to 95% (0.05). Try the following values for the degrees of freedom: 15, 10, then 5.

   Result: The critical value slowly ***increases***.

7. Finally with the probability still at 95%, try df of 20, 25, 30, and 50.

   Result: The critical value slowly ***decreases.***

Conclusion: The fewer the degrees of freedom, the higher the critical value and the more the degrees of freedom, the lower the critical value. ***This is an inverse relationship.***

[Returning to the example of first situation, when means & standard deviation given:]

Rule: *when you have a two tailed rejection region, fail to reject the null hypothesis if calculated value is greater than the critical value.*

Since the calculated value is 4.17 which ***is greater than*** the critical value of 1.997, fail to reject H_0.

CONCLUSION:

There is no significant difference between the voters in the two towns.

EXAMPLE OF SECOND SITUATION: DATA HAS BEEN COLLECTED:

The same formula for S^2 is used, but the values of S_1 and S_2 are calculated from the data.

PROBLEM:

Two universities, X and Y have similar programs in the Psychology departments. The dean of the psychology department at University X wishes to compare post-graduation salaries of their students with those of University Y to determine if there is a significant difference. Six months after graduation, the professor selected 10 graduating students from the psychology departments of each university. His hypotheses are as follows:

$H_0: \mu_X \geq \mu_Y \rightarrow$ X's post-graduate salaries $\geq$ Y's post-graduate salaries.

$H_a: \mu_X < \mu_Y$

This is a right one-tail test because of the use of $\geq$ in the null hypothesis.

RULE FOR THE RIGHT-TAILED REJECTION REGION:

When you have a *right-tailed rejection region,* fail to reject the null hypothesis if the *calculated value > critical value.*

df is 10+10-2=18

CALCULATION:

In this example, n is the same for both X and for Y and the observations for both sets, involve large interval data. ***Note: it is not necessary to have the same number of observations for both sets of data.***

STEP 1:

Calculate X and Y then add all the observations (scores) and divide by the number of observations.

$\overline{X}$ = $\sum X \div n$

= 17600 ÷ *10*

= 1760

$\overline{Y}$ = $\sum Y \div n$

= 19500 ÷ 10

= 1950

X	Y
2000	1900
1850	1800
1900	1850
2100	2000
2500	2100
2200	2300
1950	1900
1800	1700
1500	1950
2000	2000
n = 10	n = 10
$\sum X$ = 17600	$\sum Y$ = 19500
$\overline{X}$ = 1760	$\overline{Y}$ = 1950

Figure 66a: Unit 4C-5, Step 1, the collected data.

STEP 2:

X	x^2	Y	Y^2
2000	4000000	1900	3610000
1850	3422500	1800	3240000
1900	3610000	1850	3422500
2100	4410000	2000	4000000
2500	6250000	2100	4410000
2200	4840000	2300	5290000
1950	3802500	1900	3610000
1800	3240000	1700	2890000
1500	2250000	1950	3802500
2000	4000000	2000	4000000
n = 10		n=10	
$\sum X$ = 17600		$\sum Y$ = 19500	
$(\sum x)^2$ = 309760000	$\sum x^2$ = 39525000	$(\sum Y)^2$ = 380250000	$\sum Y^2$ = 38275000

Figure 66b: Unit 4C-5, Step 2, calculation.

It is necessary to calculate the variance for each set of data Sx_2 and Sy_2. Then use the individual variances to calculate the total combined variance ST_2.

$$S_X^2 = \frac{1}{n}\left[\Sigma X^2 - \frac{(\Sigma X)^2}{n}\right]$$

> = 1/10[(39525000) − (309760000 ÷ 10)
> = 0.1[39525000-30976000]
> = 0.1[8549000]
> = 854900

$$S_Y^2 = \frac{1}{n}\left[\Sigma Y^2 - \frac{(\Sigma Y)^2}{n}\right]$$

> = 0.1[38275000-(380250000 ÷ 10)]
> = 0.1[38275000-38025000]
> = 0.1[250000]
> = 25000

$$S_T^2 = \frac{n_1 s_1^2 + n_2 s_2^2}{n_1 + n_2 - 2}$$

> = [10(854900) + 10(250000)] ÷ ([10 + 10 -2]
> = [8549000 + (2500000)] ÷ (18)
> = 11049000 ÷ 18
> = 613833 to the nearest unit

STEP 3:

$$t = \frac{\overline{X} - \overline{Y}}{\sqrt{S_T^2\left[\frac{1}{n_1} + \frac{1}{n_2}\right]}}$$

> = 1760 − 1950 ÷ [√(61393[1/10 + 1/10]
> = 1760 − 1950 ÷ [√(61393[0.1 + 0.1)]
> = -190 ÷ [√(61393[0.2)]
> = -190 ÷ [√12278.6]
> = -190 ÷ [110.8]
> = -1.7148

Step 4:

Calculated t   = |-1.7148|
             = 1.7148

α = 0.05; right tail

df is 10+10-2 = 18. The calculator below, in this example, is used to produce the critical t-value.

If not already there, go to the web site and click on the second calculator: Student's t-distribution.

http://www.tutor-homework.com/statistics_tables/statistics_tables.html

To use this Student's t-distribution you need to complete the steps 1 to 4 inclusive.

1. Select the type of data distribution (right tail).

2. Write in the value for the degrees of freedom (18).

3. Write in the probability (0.05).

4. Select the direction arrow to start the calculation.

Critical t = 1.734

Calculated t = 1.7148 → Calculated t < critical t

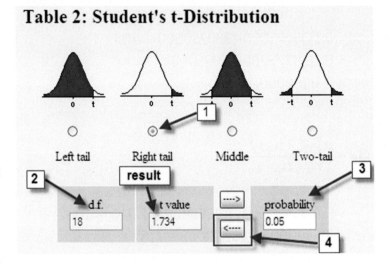

Figure 66c: Unit 4C-5, Step 4, Determining critical t.

Since the calculated value is less than the critical value, reject the null hypothesis.

CONCLUSION:

H_a: μ_X < μ_Y → accept the alternative hypothesis.

Post-graduate salaries of University X are less than post-graduate salaries of University Y.

THIRD SITUATION: USING ESTIMATED MEANS:

This is where you have two groups of data and need to calculate the means of each group and the variances. Instead of calculating the means, estimate them.

EXAMPLE 3:

The manufacturer of cups of yogurt has just added a second flavor to their lineup (original was strawberry, the new flavor is raspberry). The sugar content in milligrams of cups of yogurt from the two lines being processed was tested. Is the sugar content the same for both lines of yogurt?

Note that eleven samples were obtained for the first line (strawberry) and twelve for the second.

$H_0: \mu_1 = \mu_2$ two-tail distribution

$H_a: \mu_1 \neq \mu_2$

X	Y
23	26
25	27
22	25
24	28
27	30
22	28
23	26
25	27
24	29
23	27
20	30
	24
n = 11	n = 12

The manufacturer's desire is for the sugar content in the new line to be the same as the original. If the null hypothesis is rejected, the recipe for the second production line must be adjusted.

df = 11 + 12 - 2 = 21

Figure 67a: 4-C5, Data for Example 3, unsorted.

It does not matter if you sort the data or not, although sorting will make it a little easier to select estimates for the means. The means you select are comparatively irrelevant.

First: we'll do the calculation with unsorted data:

X	Y
23	26
25	27
22	25
24	28
27	30
22	28
23	26
25	27
24	29
23	27
20	30
	24
n = 11	n = 12

STEP 2:

Select a value for each of the two means:

A (mean for X values) = 22

B (mean for Y values) = 26

Figure 67b: Select an estimated mean for each set of data.

STEP 3:

Calculate X - A and Y – B.

X	X-A	Y	Y-B
23	1	26	0
25	3	27	1
22	0	25	-1
24	2	28	2
27	5	30	4
22 = A	0	28	2
23	1	26 = B	0
25	3	27	1
24	2	29	3
23	1	27	1
20	-2	30	4
-	-	24	-2

Figure 67c: Unit 4C-5, Step 3, determine X - A and Y - B.

STEP 4: Calculations.

Determine the values of X – A (dx) and Y – B (dy), dx^2, dy^2, the (sum of dx)^2 and (sum of dy)^2 and the sum of dx^2 and of dy^2.

X	X–A = dx	(X–A)² = dx²	Y	Y–B = dy	(Y–B)² = dy²
23	23–22 = 1	1	26	26–26 = 0	0
25	25–22 = 3	9	27	27–26 = 1	1
22	22–22 = 0	0	25	25–26 = –1	1
24	24–22 = 2	4	28	28–26 = 2	4
27	27–22 = 5	25	30	30–26 = 4	16
22 = A	22–22 = 0	0	28	28–26 = 2	4
23	23–22 = 1	1	26 = B	26–26 = 0	0
25	25–22 = 3	9	27	27–26 = 1	1
24	24–22 = 2	4	29	29–26 = 3	9
23	23–22 = 1	1	27	27–26 = 1	1
20	20–22 = –2	4	30	30–26 = 4	16
			24	24–26 = –2	4
n_1=11	Σdx = 16		n_2=12	Σdy=15	
	$(\Sigma dx)^2$ = 256	$\Sigma(dx^2)$= 58		$(\Sigma dy)^2$ = 225	$\Sigma(dy^2)$= 57

Figure 67d: Unit 4C-5, Step 3, showing the calculations.

SUMMARY: of the calculations from Figure 67d.

$3dx^2$ = 58

$(3dx)^2$ = 256

$3dy^2$ = 62

$(3dy)^2$ = 186

n_1 = 11
n_2 = 12

df = 11 +12 -2 = 21

$$S_T^2 = \frac{1}{n_1+n_2-2}\left\{\left[\sum dx^2 - \frac{(\sum dx)^2}{n_1}\right] + \left[\sum dy^2 - \frac{(\sum dy)^2}{n_2}\right]\right\}$$

= [1 ÷ (11 + 12 − 2) * {[58 − (256 ÷ 11)] + [62 − (186 ÷ 12)]}

= [1 ÷ (21) * {[58 − 23.27] + [62 − 15.5]}

= 0.0476 * {34.73- 46.5}

= 0.0476 * {-11.77}

= 0.56   to 2 decimal places

$$t = \frac{A - B}{\sqrt{S_T^2\left[\dfrac{1}{n_1}+\dfrac{1}{n_2}\right]}}$$

= (22 − 26) ÷ √[(0.56) * (1/11 + 1/12)]

= (-4) ÷ √[(0.56) * (0.090 + (0.083)]

= (-4) ÷ √[(0.56) * (0.173)]

= (-4) ÷ √[0.09688]

= (-4) ÷ [0.3113]

= -12.85   to 2 decimal places

Calculated t = |-12.86| = 12.85

Critical t = 2.08

Calculated t > critical t.

Critical t-value = 2.08

Calculated value = 12.85

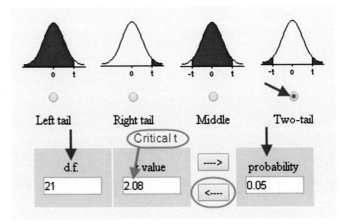

Figure 67e: Unit 4C-5, t-test, critical t.

Calculated value > critical value **0** fail to reject H_0

Conclusion: The recipe for the second automatic line doesn't need to be adjusted.

UNIT 4D: CORRELATION TESTS - ANALYZING INTERVAL & ORDINAL DATA

PURPOSE:

In the z- and t-tests for means and proportions the tests involved one or two variables and one important aspect of these tests was that they all tested interval data. Tests for data in the interval range are referred to as **parametric tests.** Tests for ordinal (ranked data) and nominal (categories or labels) are called **non-parametric tests**.

There is another test which may be used for two variables, that of a **relationship** known as **correlation**. Correlation test are used to determine if there is a relationship between two variables. The correlation tests are perhaps unusual in that one of the tests is used on parametric or interval data (Pearson's Product Moment Correlation); the other (Spearman Rank Order Correlation) tests non-parametric data, that is, ordinal data.

For the **_Pearson's Product Moment Coefficient of Correlation_** the observed numeric measures of variables (interval data) are used in the calculation. We use Pearson's test to determine if there is a correlation between the datasets, **X** and **Y**. The data may be simple or complex and the type of data requires the use of different formulas. Data must be paired, that is, for every value for **X** there must be a corresponding value for **Y**, so that the number of observations in each dataset will be the same. The letter **r** denotes the **correlation coefficient**. The value of **r**, by definition, ranges between -1 and +1. The range between negative 1 and zero is referred to a **negative** correlation. The range from zero to positive 1 is referred to as a **positive** correlation. The closer **r** is to either +1 or -1, the greater the correlation, that is the more closely the two variables are related.

The Spearman **_Rank Order Correlation_** deals with ranked data. Note that the data may be originally ranked data but it can also be interval datasets with the data unranked and the rankings are then supplied and used for Spearman's correlation test.

OBJECTIVES:

- Define the concept of correlation including a description of what information correlation tests will provide.

- Define the concepts of positive (direct), negative (inverse) and no correlation.

- Describe the major difference between the Pearson Product Moment Correlation Coefficient and the Spearman's Rank Order Correlation Coefficient.

- Calculate the Pearson's coefficient of correlation using interval data.

- Calculate the Spearman's rank order correlation coefficient using ordinal data.

INTRODUCTION:

Correlation is a statistical concept which describes the relationship between two variables in numeric terms. These variables may be interval data such as height, age, weight, size of shoe, number of pages in a book, a student's year in school, etc. Or they may be ordinal data such as rankings in an undergraduate class - comparing whether Jane's ranking in Math is similar to her ranking in English; or comparing the rankings of two classes taught the same concepts by the same professor to determine if the rankings are similar or different. If different, the researcher would want to know why. There are two forms of the correlation test: Pearson's **Product Moment Correlation Coefficient which tests parametric or interval data, and Spearman Ranked Order Correlation which tests non-parametric or ranked data.**

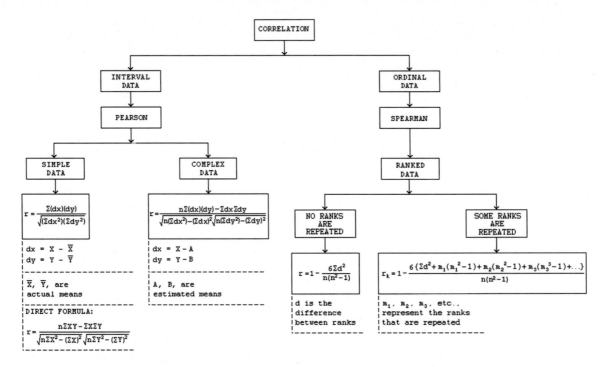

Figure 68: the analysis paths for Pearson's and for Spearman's correlation.

Simple data are small values, such as 1, 2, 3, etc., where a mean is easily calculated. Complex data requires estimated means. Concepts that may interest social science researchers are the correspondence between, for example, reading and spelling. Do children who get good scores in spelling also get good scores in reading and/or vice-versa? If a child's reading score is low, does that mean that the spelling scores will also be low?

Another example for checking the correlation between subjects is the ACT which can be used to compare the relationship between a student's Math and English skills. Do high Math scores also correlate with high scores in English? On the other hand, how many examples would show that people's Math scores have no correlation with their English score? Other concepts which involve ranked data could include ranking in exam scores, rankings in Olympian winter contests, or the (infamous?) BCS football rankings, etc.

Now some rankings may or may not be repeated. An example of repetition would be when two people tie for second place meaning that the rankings are 1, 2, 2, 4, 5, etc. If there is a large field of contestants, several ties may occur, such as 1, 2, 2, 4, 5, 6, 6, 6, 9, and so on. Notice that following a tie the rank which would normally occur is left out. The ties for 2 and 2 occupy the "space" for 3 as well as 2. Similarly, the tied data for 6 occupy the "spaces" for 7, and 8. Clearly, data where there are repeated rankings must be treated differently than data where there are no repetitious rankings.

HOW DOES ONE DEFINE A RELATIONSHIP?

Here are three relationships, common in the correlation of interval data.

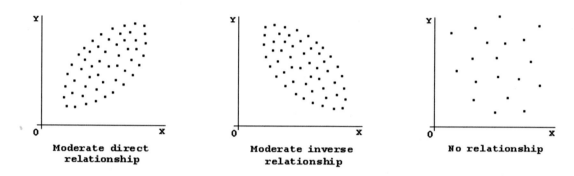

Figure 69: Data plots showing the three major correlation relationships: positive, negative and none (or negligible).

A positive or direct relationship shows an inclination for the data to "lean" to the right. A negative relation shows a tendency to "lean" to the left. Where data shows no relationship there is no pattern. The moderate direct relationship is also referred to as a positive relationship. The trend of the data points is positive. The middle graphic shows an inverse relationship with the data points trending in a negative direction. The graphic on the right shows a scattering of the data points with no trend at all.

Let's consider some examples: One pair of variables that often show a relationship is that of height and weight. In general, the taller a person is, the greater their weight. But the correlation may not always show a perfect relationship because we have some people who are excessively heavy (obese) who may not be all that tall. We also have very tall people who are excessively thin (cadaverous). But when measuring those who may be considered to be "normal" in height and weight, the height/weight relationship generally holds true. ***Note:*** If our dataset showing height/weight relationship includes either or both obese and cadaverous examples, we would have **gross outliers** which would ***skew*** our data.

The **Spearman rank order correlation** shows relationships between two different sets of rankings, such as the order in which the same set of contestants come in the 100 meter and the 400 meter races. Or compare the ranking of a person's score in the midterm with their score in the final exam or between two different subject areas such as IQ scores and performance in science subjects.

UNIT 4D-1: PEARSON'S PRODUCT MOMENT CORRELATION

PURPOSE:

There are several different types of correlation that can be shown by Pearson's test. For example, there are positive and inverse relationships. When considering the ocean, the greater the depth measured, the greater the water pressure. Conversely, as the water pressure increases, so does the depth below sea level, *a positive relationship,* since as one factor increases so does the other factor.

Two datasets may have an **inverse relationship**, that is, as one measure ***increases***, the other ***decreases***. Suppose you are looking at the temperature of the atmosphere at different heights above sea level, a general rule is that the greater the height above sea level, the lower the temperature measured. This is an inverse relationship as height ***increases*** at the same time temperature ***decreases***.

In determining the factors/parameters one is to study, there is another consideration. Note that the presence of a relationship does not guarantee a cause-effect relationship. For example, if height and weight are correlated, we cannot claim that an increase in height ***causes*** an increase in weight, rather it is ***associated*** **with** an increase in weight. And we know that an increase in weight among adults does **not** cause an increase in height, though some may wish it did! All we can say is that there seems to be a relationship" because of which, an increase in one variable is accompanied by an increase in the other variable. Or, in the case of a negative correlation, the increase in one variable is accompanied by a decrease in the other.

OBJECTIVES:

- Calculate the coefficient of correlation first using simple data then using complex data.

- Given sets of data, X and Y with the number of observations in each set being the same, determine the degree of correlation between each set.

- Calculate the Pearson's coefficient of correlation using interval data.

INTRODUCTION:

In research studies, one major type of correlation is the **Pearson Product Moment Correlation Coefficient** which measures the degree of dependence between two sets of variables. There are two forms of the Pearson test used on interval data. Simple data (1, 2, 3, 4, 5, etc.) has a simpler formula which uses actual means. There is also a "direct" formula using the variables for **X** and **Y**. Complex data requires a more complex formula and uses estimated means.

As mentioned before, the coefficient of correlation can be any number between the value -1 and +1; that is, the coefficient of correlation cannot be less than -1 nor can it be greater than +1. Usually the range is divided into three parts. These parts, however, are not absolute, only approximate.

There are two paths for the Pearson Product Moment Correlation test:

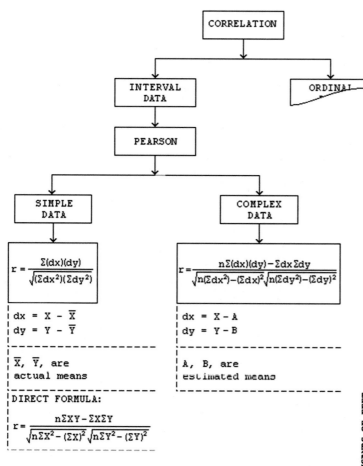

The path for simple data compares two datasets, **X** and **Y**, where actual means can be calculated.

The path for complex data uses estimated means, **A** and **B**.

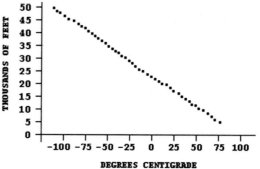

Figure 70: the two correlation

Figure 72b: Height correlated with temperature above sea level.

POSITIVE & INVERSE RELATIONSHIPS:

The correlation coefficient, **r**, is, by definition, equal to a range of values between +1 and -1. When there is no correlation at all or a negligible correlation, the value of **r** is **zero or close to zero**. **A positive correlation**, zero through positive one, shows a relationship where both variables increase. **A negative correlation,** zero through negative one shows an **inverse** relationship. The in between segments of the range are moderate negative or moderate positive relationships.

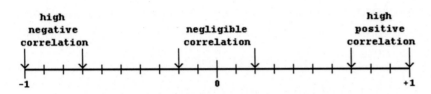

Figure 71 Areas of *r* with negative, negligible, & positive correlations.

Using the term "by definition" makes a statement of the **exact** meaning of a word or concept; that is, the value of r has a fixed range.

The closer **r** is to either +1 or -1 the greater the correlation. When the correlation is negative, we are usually only interested in a high negative relationship (between -0.75 and -1.0).

-1.0 to -0.75   →   High negative correlation

-0.75 to 0.5   →   Moderate negative correlation

-0.5 to zero   →   Low to none negative correlation

Zero to 0.5   →   Low to none positive correlation

0.5 to 0.75   →   Moderate positive correlation

0.75 to 1.0   →   High positive correlation.

It is important to understand the relationship that may be studied. For example, we can observe a relationship between the size of type used in a book and the number of words per page. We can safely predict that the larger the size of type, the fewer the number of words on the page, an inverse relationship.

With orchestral wind instruments: as a general rule, the smaller the instrument (diameter and length) the higher the pitch of the notes. For example, small wind instruments have a higher pitch than large ones. The flute has a higher pitch than the clarinet. A piccolo, which is a small flute, sounds an octave higher than an ordinary flute. Note that, with regard to wind instruments, there are three variables: diameter, length and pitch, so the relationship is more complicated. As the diameter decreases, the length also decreases, a positive relationship, but the smaller the diameter or the shorter the length, the higher the pitch, an inverse relationship.

DETERMINING WHAT RELATIONSHIP TO STUDY:

It is important to define the relationship accurately. If we were looking at books being purchased at the university bookstore, it wouldn't make any sense to check to see if there was a relationship between the first name of the person buying a book and the number of books purchased. It might make sense to see if there is a relationship between the major of a student and the categories of books that student purchases.

A question could be asked about the relationship between a region's rainfall amounts and the width of the rings showing annual growth of trees in that region. For example, in wet years the tree rings for most trees in the region are wider, in drought years the rings are narrower. One might want to use the width of rings to determine the past rainfall history of the region being studied. This relationship is a positive one: the more rain the wider the ring, the less rain the narrower the ring. But one would have to be careful to determine the parameters of the study.

First, for an analysis of the correlation we might be wise to confine the samples we study to the same type of tree. Second, we should also consider the variations in the region for our study. Suppose the trees in the study are growing on the side of a mountain. One should ask the question: are we studying the side where the prevailing winds are blowing? This would produce heavier rainfall than the rain shadow side of the mountain which has decreased rainfall. One would expect the tree rings on the rain shadow side to be narrower than on the rainy side. One might hypothesize that even though there is less rain in the wind shadow, the rainy years would produce wider rings on that side even though they may be narrower than the rings on the rainy side.

With a study of this kind, there are other considerations, such as the botanical aspects of the trees. One could postulate that the larger the tree the wider its rings are likely to be: a redwood would have much wider rings than an aspen or birch. Comparing the rings of trees in a stand of redwoods to the rings of a different stand of aspens probably wouldn't produce good data. However, one's hypothesis could be that trees of all types would be similarly affected by the same amount of rainfall. In this case, one would need to find the range of redwood annual ring growth (perhaps 1/4" to 1/2") and compare with aspen ring growth (perhaps 1/10" to 3/10"). Then the question is: do the years when heavier rain falls produce wider rings than the years of low rainfall regardless of the type of tree? If the redwood rings show consistent patterns of ring-width should one also expect similar patterns from aspen rings even though the actual widths would be in a different measured range.

In determining the factors/parameters one is to study, there is another consideration. Note that the presence of a relationship does not guarantee a cause-effect relationship. For example, if height and weight are correlated, we cannot claim that an increase in height **causes** an increase in weight, rather it is **associated with** an increase in weight. All we can say is that there "seems to be a relationship" because of which, an increase in one variable is accompanied by an increase in the other variable. Or, in the case of a negative correlation, the increase in one variable is accompanied by a decrease in the other.

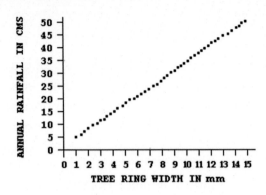

A positive correlation shows when an increase in one variable is associated with an increase in another variable. Suppose the correlation between tree ring width and rainfall was a **perfect positive correlation.** Then the correlation factor **r** would be +1. The graph of a perfect correlation would be an almost straight line rising to the right.

Figure 72a: Tree Ring Width correlated with rainfall.

A **perfect negative correlation** would be numerically represented by -1. A negative correlation occurs when an increase in one variable is associated with a decrease in the other variable such as height above sea-level and the temperature at each height measured. Of course, such perfect relationships rarely exist but real data should show a definite trend in either the positive or negative direction, that is, with the data points dropping from either right to left or left to right

[Note: the "data" in both Figure 72a and 72b is imaginary, intended merely to illustrate positive and negative trends.]

THE TWO METHODS FOR TESTING PEARSON CORRELATIONS:

There are two ways the Pearson Correlation Coefficient can be used to analyze interval data, with three different formulas. The first is the ***covariance formula*** used when the data is simple (1, 2, 3, etc.) and the sample is small.

$$r = \frac{Cov(X, Y)}{\sigma_X \sigma_Y} = \frac{\sum dxdy}{\sqrt{\sum dx^2 \sum dy^2}}$$

Cov = covariance.

dx and **dy** represent the difference between each data point and the mean.

An alternative to the covariance formula is the ***direct formula***, also used when the data is simple:

$$r = \frac{n \sum XY - \sum X \sum Y}{\sqrt{n \sum X^2 - \left(\sum X\right)^2} \sqrt{n \sum Y^2 - \left(\sum Y\right)^2}}$$

It is referred to as the "direct" formula because the actual values of **X** and **Y** are used**.**

PEARSON'S CORRELATION USING THE COVARIANCE FORMULA.

EXAMPLE 1A:

Six children were tested then during the next five days were given an opportunity to read several books. They were tested again to see if there was a correlation between the number of books read (X) and the increase in the reading scores (Y).

# of Books Read (X)	$dx = X - \overline{X}$	Increase in Reading Scores (Y)	$dy = Y - \overline{Y}$	$dxdy$
3	3-3 = 0	2	2-2.5 =-0.5	0
5	5-3 = 2	3	3-2.5 = 0.5	1
3	4-3 = 1	2	2-2.5 =-0.5	-0.5
2	2-3 =-1	0	0-2.5 =-2.5	2.5
4	4-3 = 1	4	4-2.5 = 1.5	1.5
1	1-3 =-2	4	4-2.5 = 1.5	-3
				$\sum dxdy = 1.5$

Figure 73b: Correlation - Covariance Formula, Sum of dxdy.

STEP 1:

Calculate the means for **X** and **Y**.

$$\overline{X} = \frac{3+5+4+2+4+1}{6} = \frac{19}{6} = 3.17$$

$$\overline{Y} = \frac{2+3+2+0+4+4}{6} = \frac{15}{6} = 2.5$$

[Note: to simplify the calculations, use $\overline{X}$ = 3.0]

STEP 2:

Calculate dx, dy and the sum of dxdy.

ID	Reading Score 1	# of Books Read (X)	Reading Score 2	Increase in Reading Scores (Y)
1	5	3	7	2
2	9	5	12	3
3	7	4	8	2
4	8	2	8	0
5	10	4	14	4
6	6	1	10	4
n=6				

Figure 73b: Unit 4D-1, example 1A, data.

STEP 3:

Calculate the sum of **dx²** and **dy²** (using data from **STEP 2**).

dx	dx²	dy	dy²
0	0	−0.5	0.25
2	4	0.5	0.25
1	1	−0.5	0.25
−1	1	−2.5	6.25
1	1	1.5	2.25
−2	4	1.5	2.25
	$\sum dx^2 = 11$		$\sum dy^2 = 11.5$

Figure 73c: Correlation – example 1, step 3, Sum of dx² & dy².

STEP 4:

Multiply the sum of **dx²** by the sum of **dy²** then calculate the square root of the result.

$$\sqrt{\sum dx^2 \sum dy^2} = \sqrt{11 \times 11.5} = \sqrt{126.5} = 11.25$$

STEP 5:

Find r, the correlation coefficient.

$$r = \frac{\sum dxdy}{\sqrt{\sum dx^2 \sum dy^2}} = \frac{1.5}{11.25} = 0.13$$

Conclusion: There is very little correlation between the number of books read and the increase in reading score.

PEARSON'S CORRELATION USING THE DIRECT FORMULA (SAME DATA):

EXAMPLE 1B:

Same problem: Six children were tested then during the next five days were given an opportunity to read several books. They were tested again to see if there was a correlation between the number of books read (X) and the increase in the reading scores (Y).

ID	Reading Score 1	# of Books Read (X)	Reading Score 2	Increase in Reading Scores (Y)
1	5	3	7	2
2	9	5	12	3
3	7	4	8	2
4	8	2	8	0
5	10	4	14	4
6	6	1	10	4
n=6				

Figure 74a: Unit 4D-1, example 1B, Data.

STEP 1:

Calculate the sum of X, the sum of Y and the sum of XY.

ID	# of Books Read (X)	Increase in Reading Scores (Y)	XY
1	3	2	6
2	5	3	15
3	4	2	8
4	2	0	0
5	4	4	16
6	1	4	4
n=6	$\sum X = 19$	$\sum Y = 15$	$\sum XY = 49$

Figure 74b: example 1B, sum of X, sum of Y and sum of XY.

STEP 2:

Calculate the sum of **X²**, the Square of the sum of **X**, the sum of **Y²** and the Square of the sum of **Y**.

ID	# of Books Read (X)	x^2	Increase in Reading Scores (Y)	y^2
1	3	9	2	4
2	5	25	3	9
3	4	16	2	4
4	2	4	0	0
5	4	16	4	16
6	1	1	4	16
n=6	$\left(\sum X\right)^2 = 19^2 = 361$	$\sum X^2 = 71$	$\left(\sum Y\right)^2 = 15^2 = 225$	$\sum Y^2 = 49$

Figure 74c: Square of sum X, Sum of X², Square of sum Y & Sum of Y².

STEP 3:

Calculate **r** using the direct formula:

$$r = \frac{n\sum XY - \sum X \sum Y}{\sqrt{n\sum X^2 - \left(\sum X\right)^2}\sqrt{n\sum Y^2 - \left(\sum Y\right)^2}}$$

$$= \frac{6*49 - 19*15}{\sqrt{(6*71 - (19)^2)} * \sqrt{(6*49 - (15)^2)}}$$

$$= \frac{294 - 285}{\sqrt{(426 - 361)} * \sqrt{(294 - 225)}}$$

$$= \frac{9}{\sqrt{65} * \sqrt{69}}$$

$$= \frac{9}{8.06 * 8.31}$$

$$= \frac{9}{66.98}$$

$$= 0.13$$

Both formulas give the same result, 0.13. Since this is a very low correlation, the conclusion is that there is little to no correlation between the number of books read and the increase in reading scores.

PEARSON'S CORRELATION FORMULA FOR COMPLEX DATA:

The third formula to use in correlation is for complex data (larger quantities than simple data). This formula uses estimated means.

$$r = \frac{n \sum dxdy - \sum dx \sum dy}{\sqrt{n(\sum dx^2) - (\sum dx)^2} \sqrt{n(\sum dy^2) - (\sum dy)^2}}$$

For this formula, **dx = X-A** and **dy = Y-B**, where **A** and **B** are estimated means.

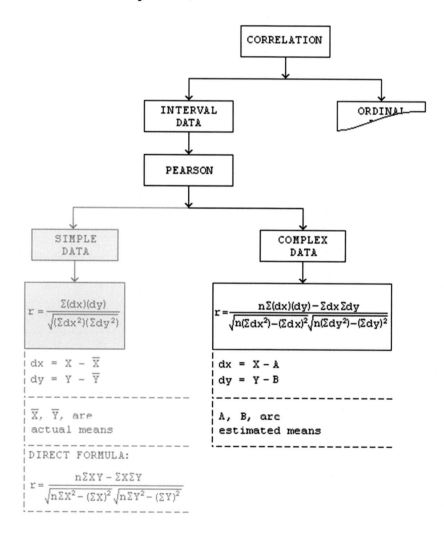

Figure 75: Emphasizing Pearson's Correlation path for complex data.

EXAMPLE 2A: USING THE FORMULA FOR COMPLEX DATA.

Age of Husband	Age of wife
23	18
27	22
28	23
29	24
30	25
31	26
33	28
35	29
36	30
39	32
n = 10	n = 10

Figure 76a: Pearson correlation for Complex Data.

STEP 1:

Choose the estimated means, **A = 30** & **B = 26**; and calculate **dx**, **dy, dx²**, **dy²** and **dxdy**.

Age of Husband (X)	dx = X-A	dx²	Age of wife (Y)	dy = X-B	dy²	dxdy
23	23-30=-7	49	18	18-26=-8	64	56
27	27-30=-3	9	22	22-26=-4	16	12
28	28-30=-2	4	23	23-26=-3	9	6
29	29-30=-1	1	24	24-26=-2	4	2
30 ← A	30-30= 0	0	25	25-26=-1	1	0
31	31-30= 1	1	26 ← B	26-26= 0	0	0
33	33-30= 3	9	28	28-26= 2	4	6
35	35-30= 5	25	29	29-26= 3	9	15
36	36-30= 6	36	30	30-26= 4	16	24
39	39-30= 9	81	32	32-26= 6	36	54
n=10	$\sum dx = 11$	$\sum dx^2 = 215$	n=10	$\sum dy = -3$	$\sum dy^2 = 159$	$\sum dxdy = 175$

Figure 76b: Unit 4D-1, Example 2a, data for husband & wife, estimated means.

STEP 2:

Calculate the coefficient of correlation.

$$r = \frac{n\sum dxdy - \sum dx \sum dy}{\sqrt{n(\sum dx^2)-(\sum dx)^2}\sqrt{n(\sum dy^2)-(\sum dy)^2}}$$

$$= \frac{(10)(175)-(11)(-3)}{\sqrt{10(215)-(11^2)}\sqrt{10(159)-(-3^2)}}$$

$$= \frac{1750 - (-33)}{\sqrt{(2150)-(121)}\sqrt{(1590)-(9)}}$$

$$= \frac{1750 + 33}{\sqrt{(2029)}\sqrt{(1581)}}$$

$$= \frac{1783}{(45.04)(39.76)}$$

$$= \frac{1783}{1790.79}$$

$$= 0.9956$$

This is a high positive correlation, the age of husband and wife are very likely to be really similar.

CAUTION: *Examine the data for husband & wife.*

The wife's age in all participant's data is lower than her husband's. In a published review of this study, a legitimate question would be: Was the selection of participants truly random? In a truly random selection one would expect some wives would be the same age or older than their husbands. *When collecting your data watch for a problem of this nature.*

EXAMPLE 2B: SECOND EXAMPLE USING THE FORMULA FOR COMPLEX DATA.

Advertizing expenses	Sales
39	47
85	53
62	58
90	86
82	62
75	68
25	60
98	91
36	51
78	84
n = 10	n = 10

An ad campaign was run in 10 different districts. Sales for each district were totaled to determine if there is a correlation between the ad expenses and subsequent sales.

Figure 77a: Unit 4D-1, example 2b, data.

STEP 1:

Choose estimated means A and B then calculate the sum of dx, sum of dx^2, sum of dy, sum of dy^2, sum of dxdy.

Advertizing expenses	Sales	dx = X-A	dx^2	dy = Y-B	dy^2	dxdy
39	47	39-82 = -43	1849	47-68 = -21	441	903
85	53	65-82 = -17	289	53-68 = -15	225	255
62	58	62-82 = -20	400	58-68 = -10	100	200
90	86	90-82 = 8	64	86-68 = 18	324	144
82 → A	62	82-82 = 0	0	62-68 = -6	36	0
75	68 → B	75-82 = -7	49	68-68 = 0	0	0
25	60	25-82 = -57	3249	60-68 = -8	64	456
98	91	98-82 = 16	256	91-68 = 23	529	368
36	51	36-82 = -46	2116	51-68 = -17	289	782
78	84	78-82 = -4	16	84-68 = 16	256	-64
n = 10	n = 10	$\sum dx$ = -170	$\sum dx^2$ = 8288	$\sum dy$ = -20	$\sum dy^2$ = 2264	$\sum dxdy$ = 3044

Figure 77b: Unit 4D-1, example 2b – Advertising & Sales, estimated means.

STEP 2:

Calculate the coefficient of correlation.

$$r = \frac{n\sum dxdy - \sum dx \sum dy}{\sqrt{n(\sum dx^2) - (\sum dx)^2}\sqrt{n(\sum dy^2) - (\sum dy)^2}}$$

$$= \frac{(10*3044) - (-170)(-20)}{\sqrt{(10*8288 - (-170)^2}\sqrt{(10* 2264) - (-20)^2}}$$

$$= \frac{30440 - 3400}{\sqrt{(82880 - 28900}\sqrt{(22640 - 400)}}$$

$$= \frac{27040}{\sqrt{(53980)}\sqrt{(22240)}}$$

$$= \frac{27040}{(232.34)(149.13)}$$

$$= \frac{27040}{34648.86}$$

$$= 0.7804$$

This gives a high correlation between the advertisement expenses and subsequent sales. The advertisement costs were a good investment.

UNIT 4D-2. SPEARMAN'S RANK ORDER CORRELATION

PURPOSE:

The other major type of correlation is the **Spearman Rank Order Correlation** which shows if there is a relationship between two different sets of rankings. The Pearson's Product Moment Correlation is based on the assumption that the relationship between variables is linear. However, it is not always possible to observe a linear relationship between variables. In such cases we can calculate Spearman's Rank Correlation which employs a different formula. As the name implies, ranks are given to the measures in each group and the calculation of the coefficient of correlation is based on these ranks. There are three situations where rankings may be used. The first and second situations

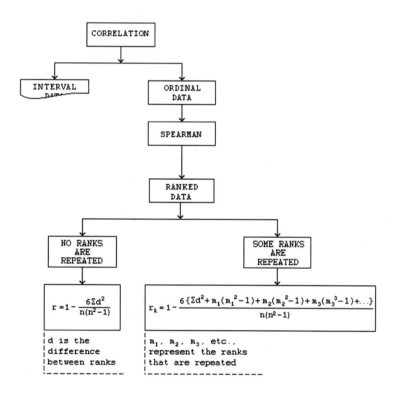

Figure 78: Analysis paths for Spearman's Ranked Data Correlation.

OBJECTIVES (from Unit 4D):

- Describe when and why you should use the Spearman and Pearson Correlation tests.

- Describe how to calculate the Pearson's coefficient of correlation and the Spearman's rank order correlation coefficient.

- Describe the major difference between the Pearson Product Moment Correlation Coefficient and the Spearman's Rank Order Correlation Coefficient.

SPEARMAN'S RANK ORDER CORRELATION:

The Pearson's Product Moment Correlation is based on the assumption that the relationship between variables is linear. However, it is not always possible to observe a linear relationship between variables. In such cases we can calculate Spearman's Rank Correlation which employs a different formula. As the name implies, ranks are given to the measures in each group and the calculation of the coefficient of correlation is based on these ranks. As with Spearman's, the numerical value of the correlation coefficient, r_s, ranges between -1 and +1. The correlation coefficient is the number indicating how the scores are relating.

In general,

If $r_s > 0$ implies positive agreement among ranks

If $r_s < 0$ implies negative agreement (or agreement in the reverse direction)

If $r_s = 0$ implies no agreement

Additionally, the closer r_s is to 1, the better the strong agreement while the closer r_s is to -1, the better is the strong agreement in the reverse direction.

There are three situations where rankings may be used. The first and second situations use the same formula:

Situation 1: Ranks are given and there are no repetitions of any rank.

Situation 2: The data is not ranked, and rankings must be supplied; no repetitions. The rankings are for the same subjects using two different variables.

Situation 3: Ranks are given and there are repetitions, for example, two different subjects are ranked third.

In the first two situations, there are no repetitions of data and the formula for **r** is:

$$r_s = 1 - \left\{ \frac{6\left(\sum D^2 \right)}{n(n^2 - 1)} \right\}$$

Where **D** is the difference between the subject's ranks on two different variables.

The ***third situation*** (ranks are given and there are repetitions) uses a modified formula with the correlation represented by r_k:

$$r_k = 1 - \frac{6\left[\sum D^2 + m_1\left(m_1^2 - 1\right) + m_2\left(m_2^2 - 1\right) + m_3\left(m_3^2 - 1\right) + \cdots\right]}{\left(n^2 - 1\right)}$$

Where m_1, m_2, m_3,... etc., represent the number of times a specific rank is repeated. For example, suppose in a dataset, the second rank (2) repeats three times, $m_1 = 3$. If, in the same dataset, the fifth rank repeats four times, then $m_2 = 4$. If there are no more repeated ranks, then m_3 is not used. (Note: the dots (...) represent additional repetitions beyond m_3.) The **k** of r_k is intended to distinguish the formula for the third situation.

If the above formulas shows a moderate to high correlation, you may wish to know how significant your result is. This student's t-distribution formula may be used to determine the significance of the result.

$$t = r\sqrt{\frac{n-2}{1-r^2}}$$

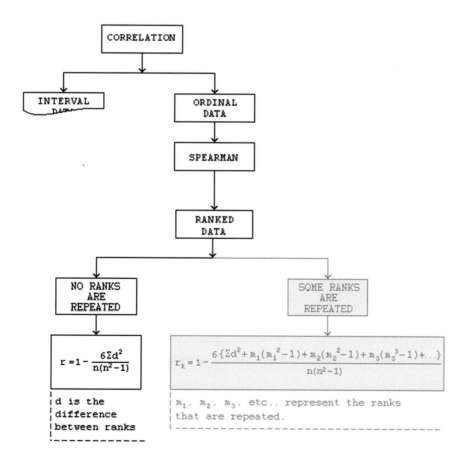

Figure 79: Emphasizing the analysis path with "no ranks are repeated."

SITUATION 1: RANKS ARE GIVEN WITH NO REPETITIONS OF ANY RANK.

Ranks are given to the same group of ten people in two different 5-mile marathons to determine if there is a correlation between their rankings.

$$D = R_1 - R_2;$$

$$n = 10$$

R_1	1	2	3	4	5	6	7	8	9	10	
R_2	1	6	8	7	10	9	3	5	2	4	
$D = R_1 - R_2$	0	-4	-5	-3	-5	-3	4	3	7	6	
D^2	0	16	25	9	25	9	16	9	49	36	$\sum D^2 = 185$

Figure 80: Unit 4D-2, Data with ranks given.

$$r_s = 1 - \left\{ \frac{6\left(\sum D^2\right)}{n(n^2 - 1)} \right\}$$

$$= 1 - \frac{6(185)}{10(100-1)}$$

$$= 1 - \frac{1110}{990}$$

$$= 1 - 1.1212$$

$$= -0.1212$$

Conclusion: While -0.1212 is negative, it is the value itself that is important. The low value of the result shows that there is little to no correlation.

Since there is little to no correspondence there is no need to use the student's t formula for significance.

SITUATION 2: THE DATA IS NOT RANKED AND RANKINGS MUST BE SUPPLIED (no repetitions).

Midterm	Final
47	39
53	84
58	62
86	90
62	82
68	55
60	85
91	98
51	46
84	96
$n = 10$	$n = 10$

When ranking must be supplied, it is based on each of the listed data sets. The rankings are for the same subjects using two different variables, for example, suppose one wanted to compare the rankings between the midterm and the final for a particular group of ten students. Here is the data **with no repetitions of ranks.**

Notice that some of the students have increased their score and others' scores have reduced.

Figure 81a: Unit 4D-2, Situation 2 Data.

STEP 1:

Sort the data.

Note: The data is paired, that is, the data for the final is matched to the data for the midterm **on the same line**. So when you sort the midterm data, you must bring the matching final data with it. This means that if the rankings are based on the midterm results, the finals will not necessarily be ranked in the same way.

STEP 2:

Supply rankings.

Midterm	Final
91	98
86	90
86	90
84	96
68	55
62	82
60	85
58	62
53	84
51	46
n = 10	n = 10

Figure 81b: Situation 2 - data sorted based on midterm data.

Midterm	Midterm Rankings R_1	Final	Final Rankings R_2
91	1	98	1
86	2	90	4
86	3	91	3
84	4	96	2
68	5	55	9
62	6	82	7
60	7	85	5
58	8	62	8
53	9	84	6
51	10	46	10
n = 10		n = 10	

Figure 81c: Columns for R1 and R2 (the rankings, highest being 1), are inserted and filled.

STEP 3:

The next step is to calculate **D**, the difference between ranks, and the square of **D**, or **D²**. Then find the sum of all **D²** or $\sum D^2$.

Figure 81d: Each D (R1-R2), D² & the sum of D2 are calculated.

Midterm	Midterm Rankings R_1	Final	Final Rankings R_2	$D = R_1 - R_2$	D^2
91	1	98	1	0	0
86	2	90	4	-2	4
86	3	91	3	1	1
84	4	96	2	1	1
68	5	55	9	-4	16
62	6	82	7	-1	1
60	7	85	5	2	4
58	8	62	8	0	0
53	9	84	6	3	9
51	10	46	10	0	0
n = 10		n = 10			$\sum D^2 = 36$

$$r_s = 1 - \left\{ \frac{6\left(\sum D^2\right)}{n(n^2-1)} \right\}$$

$$= 1 - \frac{6(32)}{10(100-1)}$$

$$= 1 - \frac{320}{990}$$

$$= 1 - 0.32 \text{ (to 2 decimal places)}$$

$$= 0.68$$

Since r_s = 0.68, a moderately high correlation, one can use the student's t-distribution to determine the significance. Note that the df is n-2 or 8.

$$t = r\sqrt{\frac{n-2}{1-r^2}}$$

$$= 0.68\sqrt{\frac{10-2}{1-(0.68)^2}}$$

$$= 0.68\sqrt{\frac{8}{1-0.4624}}$$

$$= 0.68\sqrt{\frac{8}{0.5376}}$$

$$= 0.68\sqrt{14.881}$$

$$= 0.68\,(3.8576)$$

$$= 2.6232$$

This is a two-tailed test.

H_0: r = 0   (There is no correlation in the rankings.)

H_a: r ≠ 0   (There is a significant correlation

α = 0.01, df = 9

α df	0.500	0.400	0.300	0.200	0.100	0.050	0.020	0.010
1	1.000	1.376	1.963	3.078	6.314	12.706	31.823	63.657
2	0.816	1.061	1.386	1.866	2.290	4.303	6.965	9.925
3	0.765	0.978	1.250	1.638	2.353	3.812	4.541	5.841
4	0.741	0.941	1.190	1.533	2.132	2.776	3.747	4.604
5	0.727	0.920	1.156	1,476	2.015	2.571	3.365	4.032
6	0.718	0.906	1.134	1.440	1.943	2.477	3.143	3.707
7	0.711	0.896	1.119	1.415	1.895	2.356	2.988	3.499
8	0.706	0.889	1.108	1.397	1.860	2.306	2.896	3.355
9	0.703	0.883	1.110	1.383	1.833	2.262	2.821	3.250
10	0.700	0.879	1.093	1.372	1.812	2.228	2.764	3.169
11	0.697	0.876	1.088	1.363	1.796	2.201	2.718	3.106
		0.873	1.083	1.356	1.782	2.179	2.681	3.055

Figure 81e: t-table for df of 8 at 0.05 significance.

Critical t = 2.306

Calculated t = 1.6232

Calculated t < critical t

When you have a two-tailed rejection region, fail to reject the null hypothesis if the |calculated value| > critical value.

Since the calculated value is **less than** the critical value, reject H_0.

Conclusion: There is a moderately high positive correlation between the mid-term and final rankings at a significance level of 0.05

SITUATION 3: RANKS ARE GIVEN AND THERE ARE REPETITIONS:

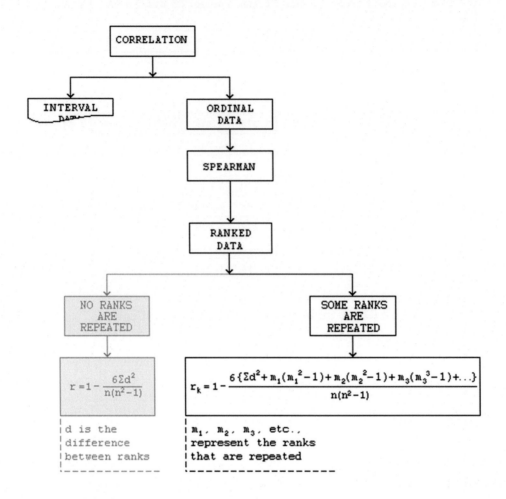

Figure 82: Emphasizing Spearman's Correlation where there are repeated ranks.

STEP 1:

Data was collected before and after training was given to 8 people to see if the training enhanced their skills. Rankings were supplied to determine the correlation between the before and after data.

IDENTIFIER	BEFORE	AFTER
A	900	1000
B	800	850
C	1000	1000
D	850	850
E	1100	1050
F	700	850
G	1000	1050
H	950	1150
n = 8		

Figure 83a: Spearman Correlation - Situation 3 unsorted data.

STEP 2:

ID	BEFORE	AFTER
F	700	850
B	800	850
D	850	850
A	900	1000
H	950	1150
C	1000	1000
G	1050	1050
E	1100	1250
n = 8		

First sort the data according to the before scores as it is easier to supply the ranks needed if the data is in order.

Because the before and after data is paired, the "after" data will not necessarily follow the same order.

Figure 83b: Spearman Correlation - situation 3 step 2, data sorted.

STEP 3:

Rank the two sets of data with 1 being the lowest value and 8 the highest.

There are no repetitions in the before ranks, there are two different repetitions in the after ranks:

$$r_k = 1 - \frac{6\left[\sum D^2 + m_1(m_1^2 - 1) + m_2(m_2^2 - 1)\right]}{n(n^2 - 1)}$$

Rank 1 → **m₁** = 3 repetitions.

Rank 4 → **m₂** = 2 repetitions.

ID	BEFORE	BEFORE RANK	AFTER	AFTER RANK
F	700	1	850	1
B	800	2	850	1
D	850	3	850	1
A	900	4	1000	4
H	950	5	1150	7
C	1000	6	1000	4
G	1050	7	1050	6
E	1100	8	1250	8
n = 8				

Figure 83c: Spearman Correlation - situation 3 step 3 ranks added.

STEP 4:

Remember that when ranks are repeated, the next rank is the number that would follow if the previous ranks had **not** been repeated. So, since rank 1 has three repetitions, the next rank is 4. Calculate d and d^2.

ID	BEFORE RANK	AFTER RANK	$D = R_1 - R_2$	D^2
F	1	1	0	0
B	2	1	1	1
D	3	1	2	4
A	4	4	0	0
H	5	7	-2	4
C	6	4	2	4
G	7	6	1	1
E	8	8	0	0
n = 8				$\sum D^2 =$ 14

Figure 83d: Spearman Correlation, Situation 3 step 4.

$$r_k = 1 - \frac{6\left[\sum D^2 + m_1(m_1{}^2 - 1) + m_2(m_2{}^2 - 1)\right]}{n(n^2 - 1)}$$

$$= 1 - \frac{6\left[14 + 3(3^2 - 1) + 2(2^2 - 1)\right]}{8(8^2 - 1)}$$

$$= 1 - \frac{6\left[14 + 24 + 6\right]}{8(63)}$$

$$= 1 - \frac{6\left[44\right]}{504}$$

$$= 1 - \frac{264}{504}$$

$$= 1 - 0.5238$$

$$= 0.4762$$

Summary:

H_0: r = 0   (There is no significant correlation.)

H_a: r ≠ 0   (There is a significant correlation at 0.05.)

α = 0.05; **df** = n-2 = 6

Rule: Two-tailed, |calculated value| > critical value then fail to reject H_0.

α = 0.05 and **df** = n-2 = 6

α df	0.500	0.400	0.300	0.200	0.100	0.050	0.020	0.010
1	1.000	1.376	1.963	3.078	6.314	12.706	31.823	63.657
2	0.816	1.061	1.386	1.866	2.290	4.303	6.965	9.925
3	0.765	0.978	1.250	1.638	2.353	3.812	4.541	5.841
4	0.741	0.941	1.190	1.533	2.132	2.776	3.747	4.604
5	0.727	0.920	1.156	1.476	2.015	2.571	3.365	4.032
6	0.718	0.906	1.134	1.440	1.943	2.477	3.143	3.707
7	0.711	0.896	1.119	1.415	1.895	2.356	2.988	3.499
8	0.706	0.889	1.108	1.397	1.860	2.306	2.896	3.355
9	0.703	0.883	1.110	1.383	1.833	2.262	2.821	3.250
10	0.700	0.879	1.093	1.372	1.812	2.228	2.764	3.169
11	0.697	0.876	1.088	1.363	1.796	2.201	2.718	3.106
		0.873	1.083	1.356	1.782	2.179	2.681	3.055

Figure 83e: Spearman Correlation t-table, df 6.

Critical value = 2.477

Calculated t = 0.4762

Since 0.4782 < 2.477; → reject H_0. There is a significant difference between the rankings.

Conclusion: The training was effective.

SUPPOSE THERE WERE DUPLICATIONS OF RANK IN BOTH BEFORE & AFTER DATA:

For example, suppose H = 900; Then the before rank for H would be 4 not 5. However, the rank of 5 is no longer used, so the only differences are circled.

STEP 1:

ID	BEFORE	BEFORE RANK	AFTER	AFTER RANK
F	700	1	850	1
B	800	2	850	1
D	850	3	850	1
A	900	4	1000	4
H	900	4	1150	7
C	1000	6	1000	4
G	1050	7	1050	6
E	1100	8	1250	8
n = 8				

Figure 84a: duplicate rank both before & after data

STEP 2: The calculation must change as shown by the highlights:

ID	BEFORE RANK	AFTER RANK	$D = R_1 - R_2$	D^2
F	1	1	0	0
B	2	1	1	1
D	3	1	2	4
A	4	4	0	0
H	4	7	-3	9
C	6	4	2	4
G	7	6	1	1
E	8	8	0	0
n = 8				$\sum D^2 =$ 19

Figure 84b: Spearman Correlation Situation 3 step 4.

STEP 3: In this case the only difference is the one rank, so the formula would be:

$$r_k = 1 - \frac{6\left[\sum D^2 + m_1(m_1^2 - 1)\right]}{n(n^2 - 1)}$$

So: $D^2 = 19$, $m_1 = 4$, $m_1^2 = 16$, **and** $n^2 - 1 = 64 - 1 = 63$

$$r_k = 1 - \frac{6[19 + 3(3^2 - 1)]}{8(8^2 - 1)}$$

$$= 1 - \frac{6[19 + 24]}{8(63)}$$

$$= 1 - \frac{6[43]}{504}$$

$$= 1 - \frac{258}{504}$$

$$= 1 - 0.5119$$

$$= 0.4881$$

STEP 4: α = 0.05 and **df** = n-2 = 6

Rule: Two-tailed, |calculated value| > critical value then fail to reject H_0.

df \ α	0.500	0.400	0.300	0.200	0.100	0.050	0.020	0.010
1	1.000	1.376	1.963	3.078	6.314	12.706	31.823	63.657
2	0.816	1.061	1.386	1.866	2.290	4.303	6.965	9.925
3	0.765	0.978	1.250	1.638	2.353	3.812	4.541	5.841
4	0.741	0.941	1.190	1.533	2.132	2.776	3.747	4.604
5	0.727	0.920	1.156	1.476	2.015	2.571	3.365	4.032
6	0.718	0.906	1.134	1.440	1.943	2.477	3.143	3.707
7	0.711	0.896	1.119	1.415	1.895	2.356	2.988	3.499
8	0.706	0.889	1.108	1.397	1.860	2.306	2.896	3.355
9	0.703	0.883	1.110	1.383	1.833	2.262	2.821	3.250
10	0.700	0.879	1.093	1.372	1.812	2.228	2.764	3.169
11	0.697	0.876	1.088	1.363	1.796	2.201	2.718	3.106
		0.873	1.083	1.356	1.782	2.179	2.681	3.055

Figure 84c: Spearman Correlation t-table df 6.

Critical value = 2.477

Calculated t = 0.4881

Since 0.4881 < 2.477; → reject H_0. There is a significant difference between the rankings.

Conclusion: The training was effective.

Note: The more duplicated ranks, the greater the difference would be.

UNIT 4D: ANALYZING INTERVAL & ORDINAL DATA: CORRELATION TESTS:

1. Describe when and why you should use the Spearman and Pearson Correlation tests.

2. Define the concepts of positive (direct), negative (inverse) and no correlation.

3. Describe how to calculate the Pearson's coefficient of correlation and the Spearman's rank order correlation coefficient.

4. When should you use the Student's t-distribution formula for the value of t?

UNIT UNIT 4E: THE CHI-SQUARE TEST.

PURPOSE:

Another test of non-parametric data is known as the Chi-Square test. The purpose of this unit is to show you when and how to use the chi-square test. The Greek symbol representing 'chi' (the upper case letter for **X**) is:

$$\chi^2$$

Note that 'chi' is pronounced 'kai' not 'chee'.

The Chi-value is used in a non-parametric test with nominal data, that is, data consisting of labels or categories. This distribution is a curve that shows a single left-tail OR a single right-tail.

The table is found at the end of other tables at the following URL:

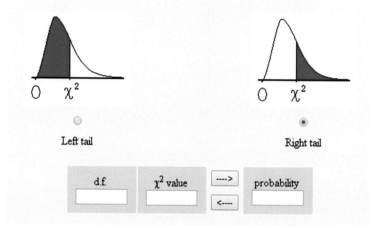

Table 3: Chi-Squared (χ^2) Distribution

Left tail

Right tail

d.f. χ^2 value   ---->   probability
                          <----

Figure 85: On line Table for Chi Square analysis.

http://www.tutor-homework. com/statistics_tables/ statistics_tables.html

ANALYSIS RULE: The null hypothesis states that χ^2 is equal to the given value and the alternative hypothesis states that the observed value will be either less than χ^2 (left tail) OR greater than χ^2 (right tail). Not both! This is a one-tailed distribution. The real question will be to determine whether the distribution is left-tailed or right-tailed.

OBJECTIVES:

1. Describe what you need to know in order to perform a Chi-Square test.

2. What is a cross-tabulation and how do you use it?

3. As soon as you have collected all your data, take your hypothesis and the nominal data you were told to collect and complete a Chi-Square test.

INTRODUCTION:

The Chi-Square test is a **non-parametric test**. When we complete a research study in the social sciences, it is not always possible to collect interval data. And even when we can, much of our data may be nominal (using names, categories, or symbols) or ordinal (ranks). Whereas the Spearman's Rank Correlation tested ordinal data, the Chi-Square tests nominal data.

THERE ARE TWO CHI-SQUARE TESTS OF MOST INTEREST:

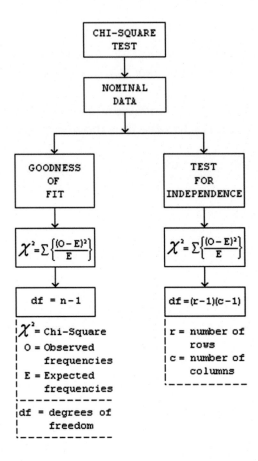

Figure 86: The flow chart for the two most commonly used Chi-Square Tests.

GOODNESS OF FIT:

This is used to determine if there is a significant difference between **E** the expected value and **O** the observed value as defined by a theory or hypothesis. This test is appropriated for situations involving one categorical variable.

INDEPENDENCE OF ATTRIBUTE:

This test is used to analyze whether two different variables are associated with each other or are independent of each other. The test uses cross-tabulations otherwise known as a contingency table. Note that the same formula is used as in the goodness of fit tests, but the layout of the data is different causing the result to be different.

DEFINITION OF THE TERM "CHI-SQUARE":

Generally Chi-Square is defined as the sum of the squares of **n** independent standard normal **variates** with n degrees of freedom.

Variate: a variable quantity that is random.

The formula for the **standard normal variate**:

$$\left\{\frac{X-\mu}{\sigma}\right\}^2$$

The **general** chi-square formula:

$$\sum_{i=1}^{n}\left\{\frac{X_i-\mu}{\sigma_i}\right\}^2$$

Where **i** represents the values from **1** through **n**.

THE CHI-SQUARE TEST:

The Chi-square test is probably the most widely used non-parametric test and is used under one of three conditions:

ONE GROUP CHI-SQUARE TEST or GOODNESS OF FIT:

Karl Pearson in 1900 developed a test to determine the significance of the difference between an experimental value, that is, the observed value and the expected or theoretical value described by the experimenter's theory or hypothesis.

As with the t-test, degrees of freedom are used. When testing a single set of observations for significance, the degrees of freedom are **n-1** where **n** is the sample size. The test, known as **goodness of fit,** determines how well the observations fit the theoretical values.

CROSS-TABULATED CHI-SQUARE TEST or INDEPENDENCE OF ATTRIBUTE:

In this case the data is gathered for several variables and placed in a grid based on the variables of interest and is referred to as the **Chi-Square test for independence**. Although most examples show only two variables, the cross tabulation may test more than two. The cross-tabulation is also known as a **contingency table.** The null hypothesis states that the variables are independent. The alternative hypothesis states that the variables are dependent.

COMPARING THE TWO MOST USEFUL χ^2 TESTS:

The basic method of conducting the chi-square test for both goodness of fit and independence is the same. The differences between these two tests are in the method of obtaining the expected value and in the formulation of the hypothesis and the degrees of freedom.

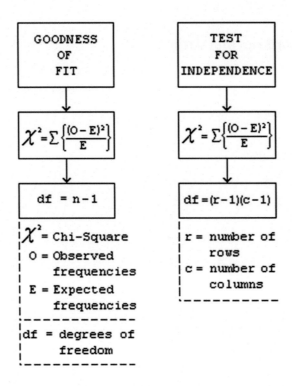

Figure 87: Comparing the two branches of the Chi-Square flow chart. Although the same formula is used, the test for independence uses a cross-tabulation matrix and the degrees of freedom are based on the number of rows and columns in the matrix.

The **goodness-of-fit test** compares the set of **observed data** with the **expected frequency** (the theoretical values for the population). In this test, it is possible to determine the values for the **expected** frequencies on the basis of the researcher's theory or hypothesis. The hypothesis in this test is that the observed and the expected frequencies are equal.

In the **test for independence** we are interested in testing the equality of the values in each cell. This test compares two or more sets of observations or variables. To make the test, we use a **cross-tabulation matrix**, where the expected frequency per cell has to be calculated by using the totals for each category.

ANALYSIS RULE:

With a one tailed rejection region, fail to reject the null hypothesis if the calculated value < critical value.

UNIT 4E-1. THE CHI-SQUARE GOODNESS OF FIT TEST

PURPOSE:

For the Pearson Goodness of fit test, nominal data is gathered and compared with the expected or predicted frequency (**E**) for each category. The null hypothesis would be that the there is no significant difference between the observed (**O**) and the expected (or predicted) frequencies.

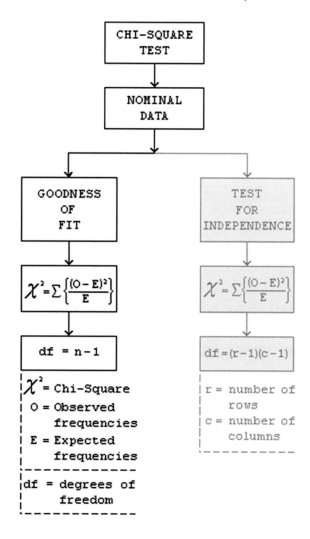

Figure 88: Emphasizing the Chi-Square Goodness of fit path.

OBJECTIVES:

- Describe and perform a Chi-Square goodness of fit test.

GOODNESS OF FIT CHI-SQUARE TEST:

This test is sometimes referred to as the test for homogeneity, meaning that all elements are the same, that is, we determine if observed frequencies (data distribution or numbers per category) match the expected frequency. The null hypothesis would be 'There is no significant difference between the observed and the predicted frequencies per cell. The formula for the test:

$$\chi^2 = \frac{\sum (O-E)^2}{E}$$

Where **O** represents the observed frequencies, that is, the frequencies collected in your gathering of data; and the **E** represents the expected frequencies, that is, the frequencies expected of the population or those dictated by your hypothesis.

The Chi-Square table is unique to the Chi-Square concepts. Critical values are found in this table based on degrees of freedom (df).

Suppose you are testing a particular group of students to determine if their I.Q.'s are the same as the population, then E = 100. The "expectation" is that all subjects would conform to the average population I.Q.

Now, whether the observed I.Q.'s would all turn out to be the average of 100, would depend on a lot of factors. There have been sufficient tests of populations to show that the average of 100 is in fact a real average for the population at large, but small to large groups when tested may show different results. For example, students attending university may show a higher IQ because fewer "average" or "below average" students are likely to attend university. This would skew the results to the right resulting in greater I.Q. scores. On the other hand, suppose you were testing all the middle school students in a particular school district who were having difficulty with their reading or math, you might find that their average I.Q. score was less than 100.

For a goodness-of-fit test you want there to be no difference between the observed scores and the expected scores. If the scores are significantly different, then the observed data does not **fit well** with the expected scores.

Chi-Square Table

one-tailed	0.050	0.010	0.001
df	95%	99%	99.9%
1	3.84146	6.63490	10.828
2	5.99147	9.21034	13.816
3	7.81473	11.3449	16.266
4	9.48773	13.2767	18.467
5	11.0705	15.0863	20.515
6	12.5916	16.8119	22.458
7	14.0671	18.4753	24.322
8	15.5073	20.0902	26.125
9	16.9190	21.6660	27.877
10	18.3070	23.2093	29.588
11	19.6751	24.7250	31.264
12	21.0261	26.2170	32.909
13	22.3621	27.6883	34.528
14	23.6848	29.1413	36.123
15	24.9958	30.5779	37.697
16	26.2962	31.9999	39.252
17	27.5871	33.4087	40.790
18	28.8693	34.8053	42.312
19	30.1435	36.1908	43.820
20	31.4104	37.5662	45.315
21	32.6705	38.9321	46.797
22	33.9244	40.2894	48.268
23	35.1725	41.6384	49.725
24	36.4151	42.9798	51.179
25	37.6525	44.3141	52.620
26	38.8852	45.6417	54.052
27	40.1133	46.9630	55.476
28	41.3372	48.2792	56.892
29	42.5569	49.5879	58.302
30	43.7729	50.8922	59.703
40	55.7585	63.6907	73.402
50	67.5048	76.1539	86.661
60	79.0819	88.3794	99.607
70	90.5312	100.425	112.317
80	101.879	112.329	124.839
90	113.145	124.116	137.208
100	124.342	135.807	149.449

Figure 89: The Chi-Square table.

EXAMPLE 1:

One thousand ten-digit telephone numbers (10,000 digits in all) were randomly selected from a directory. Then the numbers were analyzed to determine how often each digit from zero to nine appeared in these numbers.

The following data shows the observed frequency **(O)** of each digit in the selected directory numbers. If the digits occurred with equal frequency, the expected frequency **(E)** would be 1000. Test whether the randomly selected digits occur with equal frequency.

Digits	frequency (O)	(E)	(O-E)	(O-E)^2	(O-E)^2/E
0	1026	1000	26	676	676/1000 = 0.676
1	1107	1000	107	11449	11449/1000 = 11.449
2	997	1000	-3	9	9/1000 = 0.009
3	966	1000	34	1156	1156/1000 = 1.156
4	1075	1000	75	5625	5625/1000 = 5.625
5	933	1000	-67	4489	4489/1000 = 4.489
6	1107	1000	107	11449	11449/1000 = 11.449
7	972	1000	-28	784	784/1000 = 0.784
8	964	1000	-36	1296	1296/1000 = 1.296
9	1053	1000	53	2809	2809/1000 = 2.809
	10,000	10,000			$\sum\{(O-E)^2/E\}$ = 39.742

Figure 90: Chi-Square Goodness of fit Data chart with the calculations involving O-E.

SOLUTION:

Null Hypothesis **H$_0$**: The digits in the directory occur equally frequently.

Alternate **H$_a$**: The digits do not occur with equal frequency.

α = 0.05

df = n-1 = 10-1 = 9

$$\chi^2 = \frac{\sum(O-E)^2}{E}$$

= 39.742

Critical value at **0.05** level of significance and **9** degrees of freedom = 16.919.

→ Calculated value = 39.742.

Calculated value > Critical value

Chi-Square Table

one-tailed df	0.050 95%	0.010 99%	0.001 99.9%
1	3.84146	6.63490	10.828
2	5.99147	9.21034	13.816
3	7.81473	11.3449	16.266
4	9.48773	13.2767	18.467
5	11.0705	15.0863	20.515
6	12.5916	16.8119	22.458
7	14.0671	18.4753	24.322
8	15.5073	20.0902	26.125
9	16.9190	21.6660	27.877
10	18.3070	23.2093	29.588
11	19.6751	24.7250	31.264
12	21.0261	26.2170	32.909
13	22.3621	27.6883	34.528

Figure 90b: chi-square table example 1, df 9.

→ Reject **H₀:** digits do not occur equally frequently in the directory.

Conclusion: There is no "goodness of fit" between the observed results and the expected results.

Even if you used the 99% or the 99.9% significance levels, **H₀:** must still be rejected.

EXAMPLE 2:

The number of accidents in a week in a certain factory was recorded. Test whether the weekly number of accidents was the same (i.e.: there is a goodness of fit).

Week	Accidents (O)
1	11
2	9
3	10
4	2
5	8
6	10
7	9
8	6
9	9
10	4
	100

Figure 91a: goodness of fit example 2 data.

H_0 = There is no significant differences in the number of accident per week.

H_a = There is a significant difference in the number of accidents per week.

Week	Accidents (O)	(E)	(O-E)	(O-E)²	(O-E)²/E
1	11	10	1	1	1/10 = 0.1
2	9	10	-1	1	1/10 = 0.1
3	10	10	0	0	0.0
4	2	10	-8	64	64/10 = 6.4
5	8	10	-2	4	16/10 = 1.6
6	10	10	0	0	0.0
7	9	10	-1	1	1/10 = 0.1
8	6	10	-4	16	16/10 = 1.6
9	9	10	-1	1	1/10 = 0.1
10	4	10	-6	36	36/10 = 3.6
	100			$\sum(O-E)^2 =$ 124	$\sum\{(O-E)^2/E\}$ = 13.6

$\alpha = 0.05$

df = n-1 = 10-1 = 9

STEP 1:

The expected number **(E)** of accidents is 10 per week.

Calculate **(O − E)²/E**:

Figure 91b: The Chi-Square Goodness of fit example 2 calculation.

SOLUTION:

$$\chi^2 = \frac{\sum (O-E)^2}{E}$$

= 13.6 (calculated value)

Critical value at **0.05**, with **9** df = 16.9190

Calculated value = 13.6

Compare the calculated value with the critical value:

13.6 < 16.9190

Calculated value < critical value

RULE: When the calculated value < critical value **0** Fail to reject **H₀**.

Chi-Square Table

one-tailed	0.050	0.010	0.001
df	95%	99%	99.9%
1	3.84146	6.63490	10.828
2	5.99147	9.21034	13.816
3	7.81473	11.3449	16.266
4	9.48773	13.2767	18.467
5	11.0705	15.0863	20.515
6	12.5916	16.8119	22.458
7	14.0671	18.4753	24.322
8	15.5073	20.0902	26.125
9	16.9190	21.6660	27.877
10	18.3070	23.2093	29.588
11	19.6751	24.7250	31.264
12	21.0261	26.2170	32.909
13	22.3621	27.6883	34.528

Figure 91c: Chi-square table, example 2 df 9.

Therefore there is no significant difference in the number of accidents per week.

→ **Conclusion:** There is a "goodness of fit" between the observed results and the expected results.

Using the following site is a bit more complicated because of having to decide if you have a left tail or a right tail:

http://www.tutor-homework.com/statistics_tables/statistics_tables.html

Select Left-tail, add the degrees of freedom (9 df), the ^{2} value and press the upper arrow.

The result is a probability greater than 86%.

Now select Right-tail, and you only need to press the upper arrow to get the recalculation.

The result is less than 14%.

Using the table is much simpler.

UNIT 4E-2. THE CHI-SQUARE TEST FOR INDEPENDENCE

PURPOSE:

For the Chi Square Test for Independence, nominal data, that is Observed data (**O**), is gathered and compared with the expected or predicted frequency (**E**) for each category. A cross-tabulation, a grid, is set up to contain the data. The null hypothesis would be that the there is no significant difference between the observed and the expected (or predicted) frequencies.

TEST FOR INDEPENDENCE:

The **Chi Square Test for Independence** compares **two variables** and uses a cross-tabulation method for setting up the test. A cross-tabulation is a table where columns and rows are used for the calculation.

The same test statistic is used as in the goodness of fit test. The difference is in the method of calculating χ^2 and in calculating the degrees of freedom (df).

H_0: The two variables are independent.

Ha: The two variables are dependent.

```
┌──────────────────┐
│   CHI-SQUARE     │
│      TEST        │
└──────────────────┘
         │
         ▼
┌──────────────────┐
│    NOMINAL       │
│     DATA         │
└──────────────────┘
```

```
┌──────────────┐        ┌──────────────┐
│  GOODNESS    │        │    TEST      │
│    OF        │        │    FOR       │
│    FIT       │        │ INDEPENDENCE │
└──────────────┘        └──────────────┘
```

$$\chi^2 = \frac{\Sigma(O - E)^2}{E}$$

$$\chi^2 = \frac{\Sigma(O - E)^2}{E}$$

$$df = n-1$$

$$df = (r-1)(c-1)$$

χ^2 = Chi-Square
O = Observed frequencies
E = Expected frequencies
- - - - - - - -
df = degrees of freedom

r = number of rows
c = number of columns

Figure 92 Emphasizing the Chi-Square Test for Independence.

OBJECTIVES:

- Describe and perform the Chi-Square Test for Independence.

CROSS-TABULATION:

To do a cross-tabulation, one creates a table to contain the data. The simplest such table is two by two, having two columns and two rows for the data and a column and a row for totals. The entries in the cells indicate how columns and rows are combined for the calculation:

	Column 1	Column 2	Total
Row 1	a	c	a + c
Row 2	b	d	b + d
Total	a + b	c + d	a + b + c + d

Figure 93a: 2x2 Cross-tab example, vertical entries.

It is irrelevant whether the location of the letters representing the data (a, b, c, and d) is as in the table above where data is entered horizontally; or as in the table below where data is entered vertically, as long as you are consistent.

	Column 1	Column 2	Total
Row 1	a	b	a + b
Row 2	c	d	c + d
Total	a + c	b + d	a + b + c + d

Figure 93b: 2x2 Cross-tab example, horizontal entries.

Compare the two figures, 93a & 93b. In 93a, the information was entered vertically so the total for **Column 1** is **a + b**. In 93b, the information was entered horizontally, so the total for **Row 1** is **a + b**.

We need to know if the cell values (frequencies) are significantly different from one another (i.e., are independent). Suppose we are working with a two by two (2 x 2) cross-tabulation (Figure 97a). The data collected are the observed values (**O**). To calculate the expected values (**E**), you need the information in Figure 93c.

The row totals become **A** & **B**; the overall total becomes **C**.

	Column 1		Column 2		Total
	O	E	O	E	
Row 1	a		c		a + c = **A**
Row 2	b		d		b + d = **B**
Total	a + b		c + d		a + b + c + d = **C**

Figure 93c: Cross-tab example, columns added for expected entries.

To make the calculation of each **E**, you take each column's total and multiply it by that row's total (**A** or **B**) then divide by the overall total **C**. So we get:

Row 1 Column 1 = **A**(a+b)/**C**   Row 1 Column 2 = **A**(c+d)/**C**
Row 2 Column 1 = **B**(a+b)/**C**   Row 2 Column 2 = **B**(c+d)/**C**

Comparing these results, you can see why the process is called "cross-tabulation."

LARGER TABULATIONS:

Which order you use in completing the cross-tab grid is determined by the order in which your data is entered and your placement of the labels for row and column. For example, suppose you have collected data for Monday through Saturday for four weeks. You would probably fill in the table, week by week. If you set the columns as in the following table, the data would be inserted in Week 1 first, Week 2, second and so on. The cross tabulation entries would look like this:

	Wk 1	Wk 2	Wk 3	Wk 4	Totals
Mon	a	g	m	s	a+g+m+s = **A**
Tues	b	h	n	t	b+h+n+t = **B**
Wed	c	i	o	u	c+i+o+u = **C**
Thurs	d	j	p	v	d+j+p+v = **D**
Fri	e	k	q	w	e+k+q+w = **E**
Sat	f	l	r	x	f+l+r+x = **F**
Totals	a+b+c+d+e+f	g+h+i+j+k+l	m+n+o+p+q+r	s+t+u+v+w+x	a+b+c+d+e+f+g+h+i+j+k+l+m+n+o+p+q+r+s+t+u+v+w+x = **G**

Figure 94a: 6x4 Cross-tab example 2.

If the days of the week are horizontal, the set-up of data would look more like the following. Notice that the combinations in the cells for the totals in Figure 94a **are not the same** as the cross-tabulation in Figure 94b because the first cross-tab is a 6 x 4 tabulation and the one below is a 4 x 6, making the calculations different.

	Mon	Tues	Wed	Thurs	Fri	Sat	Totals
Wk 1	a	b	c	d	e	f	a+b+c+d+e+f = **A**
Wk 2	g	h	i	j	k	l	g+h+i+j+k+l = **B**
Wk 3	m	n	o	p	q	r	m+n+o+p+q+r = **C**
Wk 4	s	t	u	v	w	x	s+t+u+v+w+x = **D**
Totals	a+g+m+s	b+h+n+t	c+i+o+u	d+j+p+v	e+k+q+w	f+l+r+x	a+b+c+d+e+f+g+h+ i+j+k+l+m+n+o+p+ q+r+s+t+u+v+w+x = **F**

Figure 94b: 4x6 cross-tab example 2.

The expected values (**E**) for Monday, are found by multiplying **A, B, C, & D** by the **Monday total** and then divide the result by the overall total **F**:

 Wk 1 → A(a+g+m+s)/F

 Wk 2 → B(a+g+m+s)/F

 Wk 3 → C(a+g+m+s)/F

 Wk 4 → D(a+g+m+s)/F

Then the expected values for Tuesday, are found by multiplying **A, B, C, & D** by the **Tuesday** total **(b+h+n+t)/F**, the expected values for Wednesday are found by multiplying **A, B, C, & D** by the **Wednesday** total **(c+i+o+u)/F**, and so on, until you have all six days of expected values.

THE CROSS-TAB CALCULATION:

The formula for both the chi-square test for independence and the chi-square goodness-of-fit test are the same. A cross-tab test is essentially a test for the independence of data in the cells of the rows and columns. As a result, the expected values (**E**) are calculated using the totals for each category whereas the expected values for the goodness of fit test are based on your theory or hypothesis.

$$\chi^2 = \frac{\sum (O-E)^2}{E}$$

These are the **observed values:**

	Mon	Tues	Wed	Thurs	Fri	Total
a.m.	a	b	c	d	e	a+b+c+d+e
p.m.	f	g	h	i	j	f+g+h+i+j
Total	a+f	b+g	c+h	d+i	e+j	a+b+c+d+e+f+ g+h+i+j

Figure 95a: Chi-Square Independence Chart cross-tab example.

The highlighted items above are shown below so that you can see where the information used in the calculation comes from. To calculate the expected frequencies (**E**) the formulas are shown below. In order to calculate the expected value of each cell, the total for the first row is called **A,** and the total for the second row is called **B. The "Total" column shows the how the values of A, B, and C are calculated.**

	Mon		Tues		Wed		Thurs		Fri		Total
	O	E	O	E	O	E	O	E	O	E	
a.m.	a	A(a+f)÷C	b	A(b+g)÷C	c	A(c+h)÷C	d	A(d+i)÷C	e	A(e+j)÷C	a+b+c+d+e = A
p.m.	f	B(a+f)÷C	g	B(b+g)÷C	h	B(c+h)÷C	i	B(d+i)÷C	j	B(e+j)÷C	f+g+h+i+j = B
Total	a+f		b+g		c+h		d+i		e+j		a+b+c+d+e+f+g+h+ +i+j = C

Figure 95b: Chi-Square Independence Chart cross-tab example.

To summarize, to calculate each **E**, from the original cross-tab take the grand total for a row (for **a.m. = A; p.m. = B**) and multiply by the total for **each** column, then divide by the overall grand total **(C).** Repeat for each cell.

To calculate **E(a)** multiply **A (a+b+c+d+e)** by the total in column 1, **a+f.**

To calculate **E(f)** multiply **B (f+g+h+i+j)** by the total in column 1, **a+f.**

To calculate **E(b)** multiply **A (a+b+c+d+e)** by the total on column 2, **b+g.**

To calculate **E(g)** multiply **B (f+g+h+i+j)** by the total in column 2, **b+g,** etc.

To calculate each expected value the following calculations are made:

a.m. :

$$E(a) = \frac{(a+b+c+d+e)(a+f)}{a+b+c+d+e+f+g+h+i+j}$$

$$E(b) = \frac{(a+b+c+d+e)(b+g)}{a+b+c+d+e+f+g+h+i+j}$$

$$E(c) = \frac{(a+b+c+d+e)(c+h)}{a+b+c+d+e+f+g+h+i+j}$$

$$E(d) = \frac{(a+b+c+d+e)(d+i)}{a+b+c+d+e+f+g+h+i+j}$$

$$E(e) = \frac{(a+b+c+d+e)(e+j)}{a+b+c+d+e+f+g+h+i+j}$$

p.m.:

$$E(f) = \frac{(f+g+h+i+j)(a+f)}{a+b+c+d+e+f+g+h+i+j}$$

$$E(g) = \frac{(f+g+h+i+j)(b+g)}{a+b+c+d+e+f+g+h+i+j}$$

$$E(h) = \frac{(f+g+h+i+j)(c+h)}{a+b+c+d+e+f+g+h+i+j}$$

$$E(i) = \frac{(f+g+h+i+j)(d+i)}{a+b+c+d+e+f+g+h+i+j}$$

$$E(j) = \frac{(f+g+h+i+j)(e+j)}{a+b+c+d+e+f+g+h+i+j}$$

NOTE: all the a.m. expected values are multiplied by the total **A**, and divided by the grand total **C.**

NOTE: all the p.m. expected values are also multiplied by the total **A,** and divided by the grand total **C.**

The difference is that the final item used in the multiplication is the column total which matches the column indicated by the **E(α)** where **α = a, b, c, d,** or **e** for the **a.m.** data, **or** .where **α = f, g, h, i,** or **j** for the **p.m.** data.

EXAMPLE 1:

H_0: There is no significant difference between the observed and the expected (or predicted) frequencies; therefore the frequencies are dependent.

H_a: There is a significant difference between the observed and the expected (or predicted) frequencies; therefore the frequencies are independent.

An analysis of the academic credentials of a group of 50 teachers in the local school district produced the following results:

	Bachelors	Masters	Doctorate	Post Doctorate	Total
	O	O	O	O	
Male	8	6	5	3	20
Female	12	9	5	2	30
Total	20	15	10	5	50

Figure 96a: Example 1 collected data (Observed frequencies).

STEP 1:

Set up the table for the cross-tab calculation, including a column for each of the **E**, the expected frequencies.

	Bachelors		Masters		Doctorate		Post Doctorate		Total
	O	E	O	E	O	E	O	E	
Male	8	a	6	b	5	c	3	d	a+b+c+d = 20
Female	12	e	9	f	5	g	2	h	e+f+g+h = 30
Total	a+e = 20		b+f = 15		c+g = 10		d+h = 5		a+b+c+d+e+f+g+h = 50

Figure 96b: Example 1 set-up for the cross-tab calculation.

STEP 2:

Determine the expected frequencies. [Note. The expected frequency in **Column 1, Row 1**, is for the males, and is **E(a).** The matching frequency is in **Column 1, Row 2,** for the females, and is **E(e).**]

Male:

$$E(a) = \frac{(a+b+c+d)(a+e)}{a+b+c+d+e+f+g+h}$$

$$E(b) = \frac{(a+b+c+d)(b+f)}{a+b+c+d+e+f+g+h}$$

$$E(c) = \frac{(a+b+c+d)(c+g)}{a+b+c+d+e+f+g+h}$$

$$E(d) = \frac{(a+b+c+d)(d+h)}{a+b+c+d+e+f+g+h}$$

Female:

$$E(e) = \frac{(e+f+g+h)(a+e)}{a+b+c+d+e+f+g+h}$$

$$E(f) = \frac{(e+f+g+h)(b+f)}{a+b+c+d+e+f+g+h}$$

$$E(g) = \frac{(e+f+g+h)(c+g)}{a+b+c+d+e+f+g+h}$$

$$E(h) = \frac{(e+f+g+h)(d+h)}{a+b+c+d+e+f+g+h}$$

Figure 96c: example 1 setting up the cross-tab calculation formulas.

STEP 3:

Finish the calculation:

Male:

$$E(a) = \frac{(a+b+c+d)(a+e)}{a+b+c+d+e+f+g+h} = \frac{20(20)}{50} = \frac{400}{50} = 8$$

$$E(b) = \frac{(a+b+c+d)(b+f)}{a+b+c+d+e+f+g+h} = \frac{20(15)}{50} = \frac{300}{50} = 6$$

$$E(c) = \frac{(a+b+c+d)(c+g)}{a+b+c+d+e+f+g+h} = \frac{20(10)}{50} = \frac{200}{50} = 4$$

$$E(d) = \frac{(a+b+c+d)(d+h)}{a+b+c+d+e+f+g+h} = \frac{20(5)}{50} = \frac{100}{50} = 2$$

Female:

$$E(e) = \frac{(e+f+g+h)(a+e)}{a+b+c+d+e+f+g+h} = \frac{30(20)}{50} = \frac{600}{50} = 12$$

$$E(f) = \frac{(e+f+g+h)(b+f)}{a+b+c+d+e+f+g+h} = \frac{30(15)}{50} = \frac{450}{50} = 9$$

$$E(g) = \frac{(e+f+g+h)(c+g)}{a+b+c+d+e+f+g+h} = \frac{30(10)}{50} = \frac{300}{50} = 6$$

$$E(h) = \frac{(e+f+g+h)(d+h)}{a+b+c+d+e+f+g+h} = \frac{30(5)}{50} = \frac{150}{50} = 3$$

From Figure 96b:

a+b+c+d = 20

e+f+g+h = 30

a+e = 20

b+f = 15

c+g = 10

d+h = 5

Figure 96d: Example 1 cross-tab calculation completed.

STEP 4:

The expected frequencies, **E,** are added to the original matrix with **O** = Observed frequencies and **E** = Expected frequencies.

	Bachelors		Masters		Doctorate		Post Doctorate		Total
	O	E	O	E	O	E	O	E	
Male	8	8	6	6	5	4	3	2	20
Female	12	12	9	9	5	6	2	3	30
Total	20		15		10		5		50

Figure 96e: Example 1 cross-tab matrix with expected values.

STEP 5:

O	E	O – E	(O – E)²	(O – E)²/E
8	8	0	0	0 = 0.0
6	6	0	0	0 = 0.0
5	4	1	1	1/4 = 0.25
3	2	1	1	1/2 = 0.5
12	12	0	0	0 = 0.0
9	9	0	0	0 = 0.0
5	6	–1	1	1/6 = 0.167
2	3	–1	1	1/3 = 0.333
				$\sum (O-E)^2/E = 1.25$

This information needs to be placed in a table for the calculation of Chi-Square.

Figure 96f: matrix for calculating Chi-Square.

STEP 6:

Determine the significance of the result.

α   = 0.05

df   = (r-1)(c-1)

Where (in the original cross-tab):

    r = # of rows;

    c = # of columns

df   = (2-1)(4-1)

       = (1)(3)

       = 3

If the calculated value < critical value, fail to reject H_0.

Read the critical value from the table:

Chi-Square Table

one-tailed df	0.050 95%	0.010 99%	0.001 99.9%
1	3.84146	6.63490	10.828
2	5.99147	9.21034	13.816
3	7.81473	11.3449	16.266
4	9.48773	13.2767	18.467
5	11.0705	15.0863	20.515
6	12.5916	16.8119	22.458
7	14.0671	18.4753	24.322
8	15.5073	20.0902	26.125
9	16.9190	21.6660	27.877
10	18.3070	23.2093	29.588
11	19.6751	24.7250	31.264
12	21.0261	26.2170	32.909
13	22.3621	27.6883	34.528

Figure 96g: example 1, critical value = 7.81473 at 3 df.

RULE 2 - RIGHT-TAILED REJECTION REGION:

When you have a *right-tailed rejection region,* fail to reject the null hypothesis if the *calculated value > critical value.*

Calculated χ^2 < critical value

1.25 < 7.81473

Since the calculated value is NOT greater than the critical value, reject H_0.

H_a: There is a significant difference between the observed and the expected (or predicted) frequencies; therefore the frequencies are independent.

CONCLUSION: Academic credentials for male and female teachers are independent.

EXAMPLE 2:

In a survey, participants were asked the questions:

    a.   Do you drink?

    b.   Are you in favor of a local option for sale of liquor?

Test whether the opinion of the local option is dependent on whether or not an individual drinks. The results are in the table below:

Question a / b	Yes		No		Total	
Yes	56	a	31	b	87	a + b
No	28	c	25	d	53	c + d
Total	84	a + c	56	b + d	140	a + b + c + d

Figure 97a: Example 2 cross-tab setup.

α = 0.05

df   = (r-1)(c-1)   where r = # of rows; c = # of columns

      = (2-1)(2-1)

      = 1 * 1

      = 1

Yes Votes:

$$E(a) = \frac{(a+b)(a+c)}{a+b+c+d} = \frac{87(84)}{140} = \frac{7308}{140} = 52.2$$

$$E(b) = \frac{(a+b)(b+d)}{a+b+c+d} = \frac{87(56)}{140} = \frac{4872}{140} = 34.8$$

No Votes:

$$E(c) = \frac{(c+d)(a+c)}{a+b+c+d} = \frac{53(84)}{140} = \frac{4452}{140} = 31.8$$

$$E(d) = \frac{(c+d)(b+d)}{a+b+c+d} = \frac{53(56)}{140} = \frac{2968}{140} = 21.2$$

Figure 97b: example 2 cross-tab calculation.

O	E	O – E	(O – E)²	(O – E)²/E
56	52.2	3.8	14.44	$14.44/52.2 = 0.2766$
31	34.8	–3.8	14.44	$14.44/34.8 = 0.4149$
28	31.8	–3.8	14.44	$14.44/31.8 = 0.4541$
25	21.2	3.8	14.44	$14.44/21.2 = 0.6811$
				$\sum (O-E)^2/E = 1.8269$

Figure 97c: example 2 cross-tab matrix.

$$\chi^2 = \frac{\sum (O-E)^2}{E}$$

= 1.8269

Calculated value = 1.8269

Critical value = 3.84146

Calculated value < critical value

RULE:

When you have a *one-tailed rejection region,* reject the null hypothesis if the |*calculated value*| > critical value.

Fail to reject H_0; that is responses to each question are independent.

Chi-Square Table

one-tailed df	0.050 95%	0.010 99%	0.001 99.9%
1	3.84146	6.63490	10.828
2	5.99147	9.21034	13.816
3	7.81473	11.3449	16.266
4	9.48773	13.2767	18.467
5	11.0705	15.0863	20.515
6	12.5916	16.8119	22.458
7	14.0671	18.4753	24.322
8	15.5073	20.0902	26.125
9	16.9190	21.6660	27.877
10	18.3070	23.2093	29.588
11	19.6751	24.7250	31.264
12	21.0261	26.2170	32.909
13	22.3621	27.6883	34.528

Figure 97d: table example 2, df = 1.

UNIT 4E. THE CHI-SQUARE TEST Assignment:

1. Describe what you need to know in order to perform any of the two Chi-Square tests.

2. What is a cross-tabulation and how do you use it?

3. As soon as you have collected all your data, take your hypothesis and the nominal data you were told to collect and complete a Chi-Square test choosing the test you think is most appropriate for your data.

UNIT 5A: RESEARCH PROJECT - FINAL REPORT.

PURPOSE:

Aid each student in developing and/or improving skills in writing the report required for successful completion of your research project. Note: You have been asked to keep the research study for this semester small. The report may be used as the basis for a second semester study and/or for your thesis or dissertation.

YOUR REPORT FOR THIS SEMESTER:

Research for a dissertation or thesis generally requires a more formal report than is required for this class. Your research for this semester would be a smaller study and probably will not require several chapters. But your report will follow the similar lines as that described in **Appendix G**. You may use the description of each "chapter" to create your report: Introduction, research already done and your hypotheses. Describe the method you used to complete your research. Finally, describe the results and your conclusions.

OBJECTIVES:

Your final report is to be in the form of a publishable article with a minimum of a B grade. The report itself should be no more that 5-6 double spaced pages plus an appendix and bibliography. It should include the following aspects of the work you have done this semester:

1. Describe your literature search and results in justifying the hypothesis for your project. Include at the end of the report a bibliography of the articles you have selected. Use the bibliographic style required by your department.

2. Describe how you collected your data. Include in an appendix: any survey instruments or other techniques you used to collect your data; if you used apparatus, include photographic evidence of what you used.

3. Describe the tests you used to test your hypothesis and the three types of data you collected.

4. Process your data and interpret the results.

5. Describe the methods of data representation you plan to use.

6. Write a publishable article describing your research study and defend it before your discussion group (allow a maximum of one half hour for your defense).

DATA REPRESENTATIONS: Assignment.

The last couple of discussions should focus on preparing your report.

1. Review Appendix B "DOs & DON'Ts FOR REPRESENTING DATA" for a number of ideas on how to illustrate your data.

2. In your discussion group, discuss ways to illustrate the study data each of you have collected.

3. Based on the discussion, describe how your data may be graphically represented, and why it is important to include the data representations in your final report (no more than a total of 8-10 pages of double-spaced information).

4. Demonstrate how to represent your interval data in the form of a frequency distribution table.

5. Illustrate the frequency distribution with a histogram and a frequency polygon or smooth curve.

6. Decide what other graphics will be helpful to give meaning to your report. Draw and describe them.

7. Complete your report and present a "defense" of the report to the rest of the class.

UNIT 6: APPENDICES

PURPOSE:

Five appendices have been included to give summaries of important information that will help you in analyzing your data, not only for this class, but also whenever you perform a research study. Appendix g talks about writing a dissertation. Your final report for this semester can be patterned after this, but you should substitute paragraphs for chapters.

LIST OF APPENDICES:

A: DECIDING WHICH TEST TO USE.

B: DOs & DON'Ts FOR REPRESENTING DATA.

C: CHANCE, PROBABILITY & RANDOMNESS.

D: COMBINATIONS AND PERMUTATIONS: WHAT'S THE DIFFERENCE?

E: THE PROCESS FOR WRITING DISSERTATIONS OR A THESIS.

APPENDIX A. DECIDING WHICH TEST TO USE:

PURPOSE:

The purpose of this Appendix (A) is to help you in your decision making – to enable you to choose the best tests to use in analyzing your data.

The following graphic summarizes the tests for research data discussed in this book. An important aspect of these tests is deciding which test to use. The first decision point is what type of data you have: If it is Interval Data then your choices are the z-test, or the t-test, or the Pearson Correlation Test; If your data is ordinal, that is data that can be ranked, there is only one test to use: the Spearman Correlation test; If your data is nominal, that is data that can be classified or that fits into specific, non-numerical data, then there are two Chi-Square tests that can be used, depending on what you want to achieve. The rest of the flow chart details the rest of the process for completing your research.

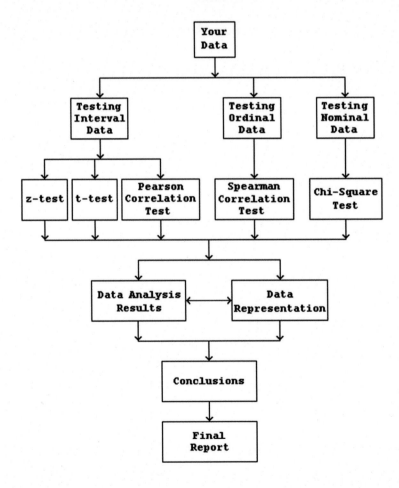

Figure 98: The decision tree for analyzing your data and completing your report.

THE MAIN PURPOSE IN ANALYZING YOUR DATA:

The main purpose is to determine if the results of your study are significant. Your null hypothesis guides your analysis, essentially saying there is no significant difference in your test results. The alternative hypothesis states what you really want to prove - that there **are** significant differences. Testing your data, using the correct method, will enable you to determine the significance of the data you have collected.

INTERVAL DATA: THE Z-TEST

There are four applications for the z-test, with either means or proportions. The z-test can analyze either one mean (one variable) or two means (two variables). If your data has more than two variables, you will need to talk to a statistic department graduate student and use the analysis of variance (ANOVAR). If you need to analyze proportions, generally, you will have one variable or two. To analyze proportions you also need to use the standard error in the calculation. Finally, to determine significance calculate the z-score and compare it with a critical value found in the z-table.

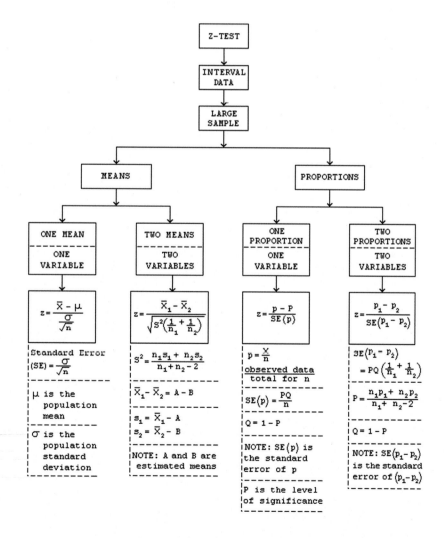

Figure 99: Decision tree for using the z-test.

DECISION INDICATORS:

1. You can compare a single mean with the population mean or a hypothesized mean. Or you can compare two means with each other; and you can use estimated means to complete the calculation. **Formulas for comparing means:**

$$z = \frac{\overline{X} - \mu}{\frac{\sigma}{\sqrt{n}}}$$

Where the population mean μ is known or an expected mean is used.

OR

$$z = \frac{\overline{X}_1 - \overline{X}_2}{\sqrt{s^2\left(\frac{1}{n_1} + \frac{1}{n_2}\right)}}$$

Where two observed means are being compared.

2. If your dataset cannot be graphed as a normal curve, by calculating the z-scores, the resulting curve shows a normal distribution.

3. A standardized set of z-scores can be used to produce grades on a curve.

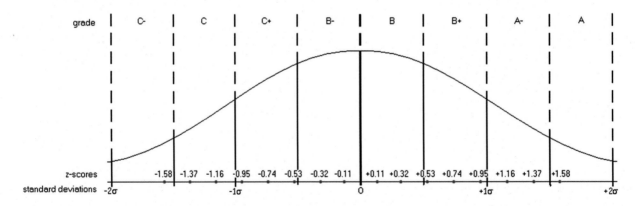

Figure 100: The z-test used for determining grades on a curve.

The following URL provides a useful method for determining significance using table 1 for the z-test:

http://www.tutor-homework.com/statistics_tables/statistics_tables.html

INTERVAL DATA: THE t-TEST:

Since the t-test is one of the most common statistical measures used in research, you will probably use it frequently. It is used with interval (normal) data, when comparing the mean for one set of data either with a hypothesized mean or with the mean for a second set of data.

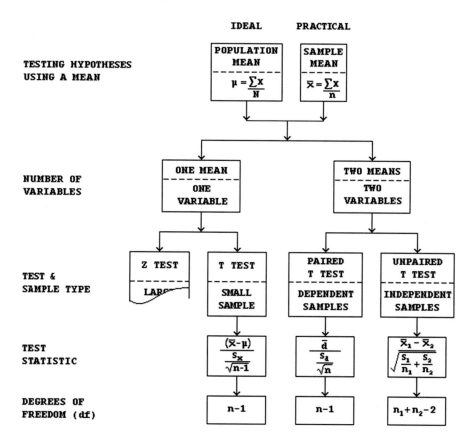

Figure 101: Three decision paths for the t-test.

DETERMINING SIGNIFICANCE:

In most t-tests, it is necessary to compare the calculated t-value with the critical t-value to determine significance. There is a t-table that can be used. Because the curve of t-values can have one or two tails, the table has two sections with the titles: "one-tail" and "two-tails." You must know whether your hypotheses show there is one tail or two.

The null hypothesis for any t-test is usually the hypothesis that your sample observations result only from chance. That means that **H_0 is always written as an equality.** This gives us three possible equalities as shown in the following statements:

- *Less than or equal to* ($\leq$) giving a one-tail test.

- *Greater than or equal to* ($\geq$) also giving a one-tail test.

- **Exactly** *equal to* (=) giving a one-tail test.

SITUATIONS WHEN THE t-TEST MAY BE USED.

The t-test may be used in any of the following situations:

a. One group, with a group mean compared with a population mean. The t-test formula:

$$t = \frac{(\bar{X} - \mu)}{\frac{S_X}{\sqrt{n-1}}}$$

b. Two groups unpaired (independent samples), comparing the mean from one group with the mean from a totally different group.

$$t = \frac{\bar{d}}{S \div \sqrt{n}}$$

c. Two groups paired (dependent samples), comparing a mean for one group's dataset with the same group's mean from a different time or a different subject.

$$t = \frac{\bar{X} - \bar{Y}}{\sqrt{S_T^2 \left[\frac{1}{n_1} + \frac{1}{n_2} \right]}}$$

DECISION INDICATORS:

SINGLE GROUP	TWO GROUP DEPENDENT (PAIRED)	TWO GROUP INDEPENDENT
Compare the mean of a data set from a single group with a population or hypothesized mean.	Compare the means of two data sets from the same subjects at different times or for different topics.	Compare the means of two data sets from two independent groups on the same topic.

Figure 102: Three criteria for using the t-test.

The test enables us to determine whether the two means we are comparing are significantly different (the alternative hypothesis). One of the means is from our random sample the other is either the mean hypothesized for the population, or the mean from a second sample. We can use the t-test provided we are dealing with either one or two groups, the data we are collecting is in the interval scale and we have a mean for each group being considered:

The t-test cannot be used if the data is nominal or ordinal. Nor can it be used when there are more than two groups being compared.

DEGREES OF FREESOM & SIGNIFICANCE:

There are two important aspects of working with t-tests. One is degrees of freedom discussed in Unit 4A-2. The other is to do with probability. In connection with the t-distribution (the probability curve), a cumulative probability is shown by an inequality, that is, it refers to the probability that a t-value or a sample mean will be less than or equal to a specified value. To determine significance the calculated t-value is compared to the critical t-value as determined by the t-table calculator.

The following URL provides a useful method for testing your data using table 2 for the t-test:

http://www.tutor-homework.com/statistics_tables/statistics_tables.html

THE CORRELATION TEST FOR INTERVAL DATA (PEARSON PRODUCT MOMENT TEST):

We use Pearson's correlation test to determine if there is a relationship between two variables, **X** and **Y**. The data may be simple or complex and use different formulas. Data must be paired, that is, for every value for **X** there must be a corresponding value for Y so that the number of observations in each dataset will be the same.

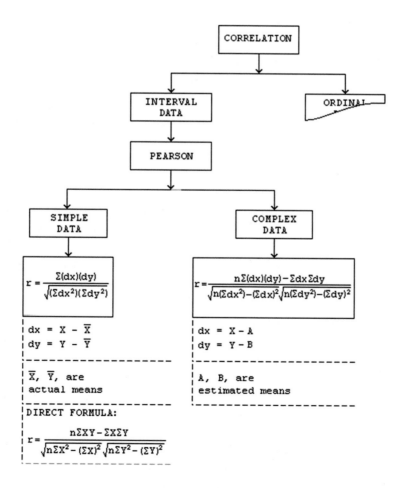

Figure 103: Two decision paths for Pearson Correlation with either simple or complex data.

RELATIONSHIPS THAT CAN BE TESTED:

Three typical correlation relationships are shown by the following graphic:

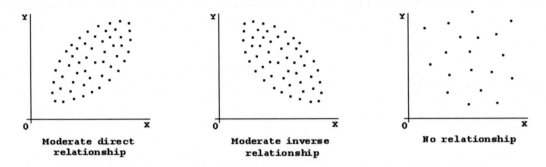

Figure 104: Three major relationships - positive, negative, & none.

The path for simple data compares two sets of data, **X** and **Y**, where actual means can be calculated. The path for complex data uses estimated means, **A** and **B**. The letter **r** denotes the **correlation coefficient**. The value of **r**, by definition, ranges between -1 and +1. The range between negative 1 and zero is referred to a **negative** correlation. The range from zero to positive 1 is referred to as a **positive** correlation. The closer **r** is to either +1 or -1, the greater the correlation, that is the more closely the two variables are related.

Figure 105: negative, negligible, positive correlation.

THE CORRELATION TEST FOR RANKED DATA (THE SPEARMAN'S RANK ORDER CORRELATION):

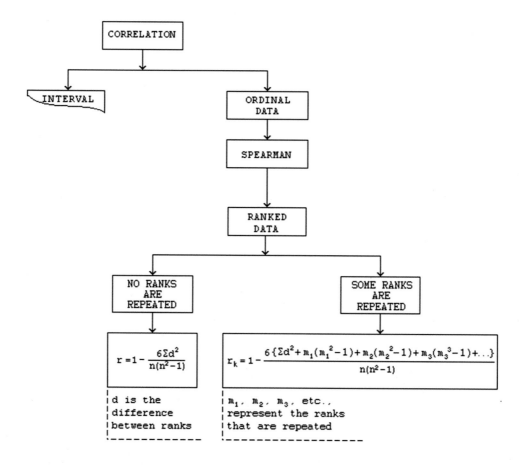

Figure 106: Two decision paths for Correlation - Pearson & Spearman.

Spearman's requires you to have two set of data, so you can make comparisons. Then the first decision concerns the type of data you have. If you have interval data you cannot use Spearman's and if you have ordinal data, you cannot use Pearson's. Of course you may have both, in which case, you may be able to use either or both tests.

The next question you need to ask is: Is the data already ranked. If not then ranks must be supplied. The next question is are any ranks repeated? There are three different conditions for the Spearman's Rank Order Correlation.

a.   Ranks are given and there are no repetitions of any rank.

$$r_s = 1 - \left\{ \frac{6\left(\sum D^2\right)}{n(n^2-1)} \right\}$$ The formula to calculate r, where $D = R_1 - R_2$

$$t = r\sqrt{\frac{n-2}{1-r^2}}$$ After the data are compared using the above formula, there is a test statistic that may be used to determine the significance of the result.

b. The data is not ranked, and rankings must be supplied; no repetitions. The rankings are for the same subjects using two different variables.

  • The same formulas are used as in condition a. The difference is the fact that rankings are not given but must be supplied.

c. Ranks are given and there are repetitions, for example, two different subjects are ranked third.

$$r_k = 1 - \frac{6\left[\sum D^2 + m_1\left(m_1^2 - 1\right) + m_2\left(m_2^2 - 1\right) + m_3\left(m_3^2 - 1\right) + \cdots\right]}{\left(n^2 - 1\right)}$$

For example, suppose rank 1 and rank 3 had repetitions, then from the above formula m_1 and m_2 are used but m_3 … are not. The dots: "…" represent any quantity of repetitions needed by the data.

THE CHI-SQUARE ('KAI'-SQUARE) TESTS:

Karl Pearson in 1900 developed a test to determine the significance of the difference between an experimental value, that is, the observed value and the expected or theoretical value described by the experimenter's theory or hypothesis for each category. There are two tests:

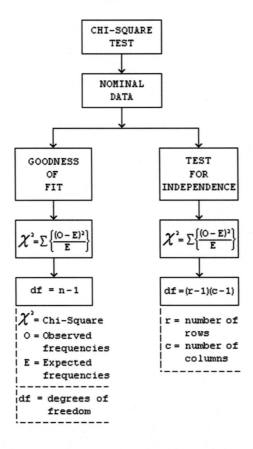

Figure 107: decision tree for Chi-Square tests.

GOODNESS OF FIT - ONE GROUP CHI-SQUARE TEST:

As with the t-test, degrees of freedom are used. When testing a single set of observations for significance, the degrees of freedom are **n-1** where **n** is the sample size. The test, known as **goodness of fit,** determines how well the observations fit the theoretical values. The null hypothesis would be that the there is no significant difference between the observed (**O**) and the expected or predicted frequencies (E).

THE INDEPENDENCE OR CROSS-TABULATED CHI-SQUARE TEST:

In this case the data is gathered for several variables and placed in a grid based on the variables of interest and is referred to as the **Chi-Square test for independence**. Although most examples show only two variables, the cross tabulation may test more than two. The cross-tabulation is also known as a **contingency table.** The null hypothesis: there are no significant relationships between the variables. The following graphic shows the simplest form of a contingency table, a two by two cross tabulation:

	Column 1	Column 2	Total
Row 1	a	b	a + b
Row 2	c	d	c + d
Total	a + c	b + d	a + b + c + d

Figure 108a: 2x2 Cross-tab example, horizontal entries.

$$\text{ROWS:} \qquad \qquad \text{COLUMNS:}$$

$$E(a) = \frac{(a+b)}{a+b+c+d} \qquad E(c) = \frac{(a+c)}{a+b+c+d}$$

$$E(b) = \frac{(c+d)}{a+b+c+d} \qquad E(d) = \frac{(b+d)}{a+b+c+d}$$

Figure 108b: formulas for calculating the expected frequencies in a 2x2 cross-tab example.

MORE ROWS AND COLUMNS:

When there are more columns and/or rows, the calculation is more complicated:

	Mon	Tues	Wed	Thurs	Fri	Total
a.m.	a	b	c	d	e	a+b+c+d+e
p.m.	f	g	h	i	j	f+g+h+i+j
Total	a+f	b+g	c+h	d+i	e+j	a+b+c+d+e+f+ g+h+i+j

Figure 108c: Cross tabulated summary shows how to prepare the matrix for calculating the expected results (E). .

Remember that **O represents the observed results (your data) and E represents the expected results determined by your theory. Here are the formula for calculating the expected values (E).**

ROWS:

$E(a) = \dfrac{(a+b+c+d)(a+e)}{a+b+c+d+e+f+g+h}$

$E(b) = \dfrac{(a+b+c+d)(b+f)}{a+b+c+d+e+f+g+h}$

$E(c) = \dfrac{(a+b+c+d)(c+g)}{a+b+c+d+e+f+g+h}$

$E(d) = \dfrac{(a+b+c+d)(d+h)}{a+b+c+d+e+f+g+h}$

COLUMNS:

$E(e) = \dfrac{(e+f+g+h)(a+e)}{a+b+c+d+e+f+g+h}$

$E(f) = \dfrac{(e+f+g+h)(b+f)}{a+b+c+d+e+f+g+h}$

$E(g) = \dfrac{(e+f+g+h)(c+g)}{a+b+c+d+e+f+g+h}$

$E(h) = \dfrac{(e+f+g+h)(d+h)}{a+b+c+d+e+f+g+h}$

Figure 108d: Cross-tab example calculation formulas.

The following URL provides a useful method for testing your data using **table 3** for the chi-square test:

http://www.tutor-homework.com/statistics_tables/statistics_tables.html

H_0 for all chi-square tests is in the form of H_0: $\alpha = m$ where m is a specific value. This means that **all chi-square tests have a single tail.**

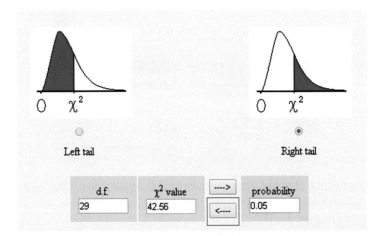

Figure 108e: the internet table for calculating Chi-Square.

The default setting is the **right tail**, but you can change it to **left tail** if you wish. Insert the df (e.g., 29) and the probability (0.05), use the left pointing value to obtain the critical chi-square value. To reject H_0, the calculated 0^2-value is compared to the critical 0^2-value as determined by the on-line table (in this case 42.56).

APPENDIX B. DOs & DON'Ts FOR REPRESENTING DATA

PURPOSE:

It is important in your report of a research study to include pictorial graphics of your data. There are many kinds of graphic representations, which one(s) you use is partly a matter of suitability and partly a matter of preference. This unit shows most of the possible graphics and describes their use. Some graphics are more useful for nominal and/or ordinal data. Others are more suitable for interval data. The most commonly used graphics are those that show differences in *frequency,* that is, for interval data.

First the researcher must collect the numeric data needed to test the hypothesis, using proper methods for collecting the data. These methods include using questionnaires or survey instruments. It is usually appropriate to include in an appendix examples of your data collection methods, your data arrays, how you arrived at your conclusions, and anything that helps to clarify your intent. Having obtained the data, it needs to be displayed in a form that can be easily interpreted. Don't forget, doing research is an effort to not only enlighten yourself but also to enlighten others through your report. Illustrations in the text of your report will enable future readers to more clearly understand what you were trying to do, how you did it, and your conclusions from the results.

The purpose in using data representations is to help make your information easier to understand and make it more meaningful to the person who reads your report. Organizing and summarizing data is often referred to as "descriptive statistics". Good and bad examples will be included in this unit.

OBJECTIVES:

- Describe how the different types of data may be graphically represented, and why it is important to include graphics, tables and other illustrations in your final report.

- Demonstrate how to represent a set of data in the form of a frequency distribution.

- Illustrate the frequency distribution with a histogram and a frequency polygon.

USING GRAPHICS TO HELP COMPARE DATA:

There are a few important cautions when considering graphic representations. Earlier in this document we used a table to show the exam scores of students taking the mid-term and the final. Suppose the professor wants to know if the students perform significantly better by the time they have completed the whole course or if the scores do not change much.

MIDTERM SCORES	# OF STUDENTS	FINAL SCORES	# OF STUDENTS
40-49	3	40-49	1
50-59	5	50-59	7
60-69	25	60-69	19
70-79	22	70-79	28
80-89	10	80-89	8
90-99	5	90-99	7
Total	70	Total	70

Figure 109: Using a table to compare data from two samples.

By drawing bar charts for each set of data, one can visually see changes. The two charts below use dotted lines. Examining them show that the scores in the second test have moved to the right. But because dots were used, it is not so easy to compare the two sets of data.

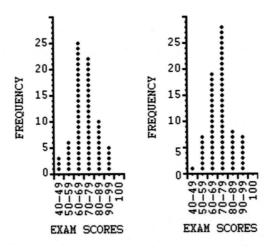

Figure 110a: a dotted bar graph has been used to illustrate the exam scores.

Perhaps by using solid bars, the information would be easier to judge?

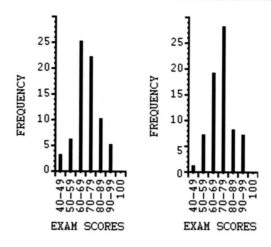

Figure 110b: solid bars are used instead of dots.

A better solution may be to combine the two sets of data into one graph. Here is one method to do that. Because two colors were used, a key is needed.

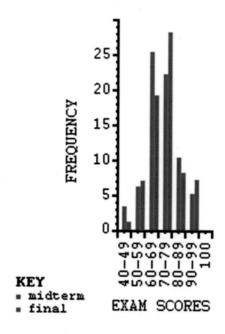

Figure 110c: if you combine two sets of data, color is helpful. But then you need a key.

When you are getting your report together, you may need to experiment in order to get a good illustration.

IMPROPER USE OF GRAPHICS:

One other aspect of representing data has not yet been considered, that is the creation of graphics that are misleading so that the reader gets the wrong impression. Another shortcoming of graphs is when the person who creates them puts insufficient information on the graphic.

The rest of Appendix B presents different type of graphics and some of the "Dos" and "DON"Ts with each type of graphic.

CONTENTS FOR APPENDIX B:

THE PIE CHART or CIRCLE GRAPHS:

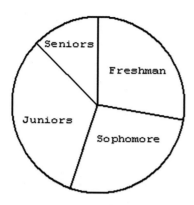
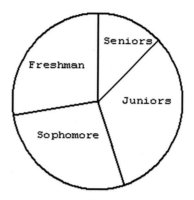

To start the graph, draw the circle and draw a line from the center to the top of the circle. Then draw the lines representing the divisions for the data.

The divisions may be drawn on the circle going to the right so it can be read in a clockwise direction. Or drawn to the left so it can be read in an anti-clockwise direction (your choice).

Figure 111a: Putting data on a pie chart.

In the U.S. we read from left to right. So the graphic which can be read clockwise (the one on the right) would probably be the best choice. The above graphic was developed to show how a circle graph is drawn. If you were using this type of graph in your report, it would be advisable to add the percentage of data falling in each segment.

But be careful: The percentages shown must sum to 100%. Degrees sum to 360° the number of degrees in a circle. Using degrees enables you to draw the pie chart, but if the graphic is intended to show percentages, the total should add up to 100%.

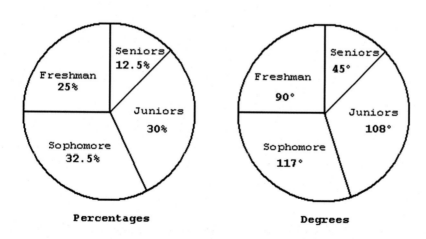

Percentages

Degrees

Figure 111b: Information added as either percentages or degrees.

NOTE: Occasionally there are some charts, where, because the percentages have been rounded, the total does not come to 100%. In the graphic on the left, if the rounding was used, 12.5% rounds to 13%, and 32.5% rounds to 33; the total would be 101%.If this occurs, mention the rounding in the caption for the graphic.

For the actual drawing of the circle graph, since one would start with the data in the form of percentages, the calculation to determine the degrees needed for each segments is the percent divided by 100 then multiplied by 360. These degree quantities are needed to draw accurate segments **0** they should **not** be used instead of percentages in labelling the graphic.

ANOTHER USE OF CIRCLES TO SHOW DATA:

Another graphic using circles to represent the data, is in the graph below that compares three department's expenditures. Using circles as the frequency of data is **not** to be recommended. Why? The problem is that the circles may be drawn so as to mislead, perhaps not intentionally. For example:

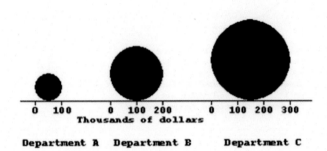

Since the human eye tends to interpret pictures visually, looking at this graphic one would tend to interpret the data as if Department C spent a great deal more than Department A when it should be only 3x Department A's spending.

Figure 112a: A graph used by a local university to report expenditures for three departments.

Assume that the graphic was drawn so that the diameter of the first circle is 1cm, the second is 2cm, the third is 3cm; the *diameters* of the circles are on the same scale as the frequency.

INTERPRETING THE GRAPHIC:

Reading the frequencies in thousands of dollars below the circles, it is fairly clear that the first circle represents $100,000 in expenditure, the second is twice as much at $200,000 and the third is three times the first at $300,000.

This problem arose because the person drawing the circles assumed that it was O.K. to just make the middle circle's diameter twice the first circle's diameter since Department B spent twice the amount as Department A; and the diameter of the third circle 3 times that of the first, since Department C spent 3 times as much as Department A; *erroneous thinking.* The **areas of the circles** are the problem. Because we tend to interpret what we see rather than what we think, we see the *areas* of the circles not the *diameters*. In thousands of dollars, the diameters of the three circles represent (A) 100, (B) 200, and (C) 300 so the radii are respectively 50, 100, and 150 (the radii are respectively 1cm, 2cm, 3cm). The real areas of these circles based on the radii: smallest = 7,900, the middle one = 31,400 or four times the area of the smallest (it should only be twice). The largest circle = 70,700, or approximately 9 times the area of the first (it should only be 3 times). A quick way of thinking about this is: since the radii are squared, the radius squared for the middle circle is two squared or 4, and the largest circle is three squared or 9.

If one were to redraw the circles so they are in **area** proportion we get this graphic.

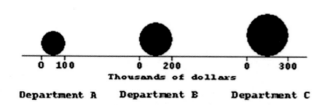

Figure 112b: Areas of the circles give more accurate visual information.

Not so dramatic! And this points up the other problem with using this type of circle. The scales for 0-100, 0-200 and 0-300 can no longer be the same (0-200 is no longer twice the length between 0 and 100, nor is 0-300 three times that length, That's why only the beginning and end points of the three scales are used here.

But, at least the comparison of the expenditures is more accurate than before.

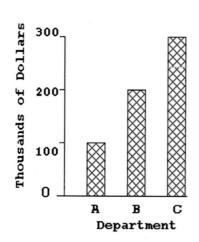

A bar graph would have best been used for this data. Here is the information in the previous circle graph drawn as a bar graph. Any kind of pattern or solid, black or color may be used to draw a bar graph. The proportion of the three bars is now clearly 1:2:3.

Figure 112c: A bar graph more accurately shows data for comparison.

MAKE SURE YOUR GRAPH CONTAINS ALL NECESSARY DATA.

Here is another bar graph which has not only insufficient information but also some essential information was left out:

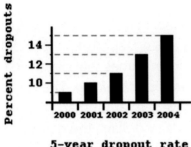

Figure 113a: Totally erroneous data representation.

This graph makes it look as if the number of dropouts in 2004 was much greater, about seven times the number of dropouts in 2000, implying that the dropout situation is significantly worse.

BUT: Notice that the **vertical axis does not begin with zero.** It actually begins with 8 with the label "8" being conveniently left out. This is a definite no-no for giving appropriate information.

Compare the faulty graph with the correctly drawn graph below and on the right. The black dashed line shows where the "8" frequencies match. By cutting off the bottom portion of the chart a totally erroneous impression is given about the dropout situation in 2004. Did someone wish to use scare tactics about dropouts? Or were they trying to make someone look bad?

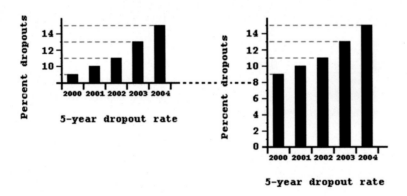

Figure 113b: comparing the erroneous chart with an accurately drawn chart.

OTHER COMMON PROBLEMS WITH DRAWING GRAPICS:

PROBLEMS OCCUR WHEN LABELING THE AXES:

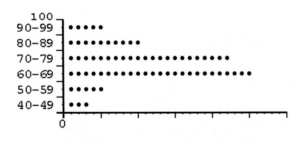

Figure 114a Bar Graph with only one axis labeled

This chart where the vertical axis is labeled but the horizontal is not is really meaningless. **QUESTIONS:** Is this data showing exam scores or test scores? Is only one person being tested or a group? Although it is implied, one cannot even be sure the age/scores increase from left to right. Without a lengthy explanation in the text, one can only guess at what this means.

This graph is poor not only because the horizontal axis is not labeled with specific ages or frequencies, but also because the number of cases represented by each data point is unknown, they may or may not be the same! Unfortunately, the reader of the report cannot guess.

Another problem that may occur is that with one of the axes, or both, although labeled, the information along each axis is less meaningful if only the "end" points of the data are included.

This graph shows what a scattergram looks like, but gives minimal details for both axes: While this might be acceptable if the goal was to show what a scattergram looked like, it would NOT be acceptable in a research study.

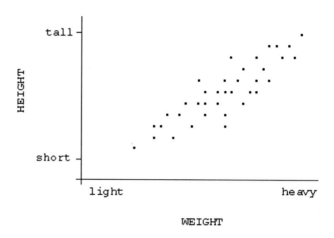

Figure 114b poorly labeled Scattergram.

ALWAYS MAKE SURE YOU PROPERLY LABEL BOTH AXES!

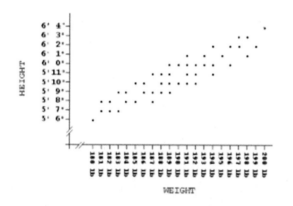

Figure 114c: Hash marks used to truncate unneeded portion(s) of axes.

Now if we are talking about, for example, the high school's basketball team and labeled the axes, the graph becomes more meaningful. Note that this type of diagram has two variables, both recording interval data, that is, height and weight.

Note the hash marks // (not hashtags #) at the beginning of both axes. These answer the question: What does one do about data that does not need to be shown since there are no representatives for that data? For example, there are no members with heights less than 5'6"; or weights less than 180 lb.

To deal with data points that are not needed, hash marks (//) may be placed on each axis to show that you have truncated the axes. Missing unneeded labels on one or both axes is acceptable if you use the hash marks. **Hash marks cannot be placed in the middle of a set of labels even though there may be no data around that point; place only at the beginning of an axis.**

When creating a graph which has two variables, think about the labels:

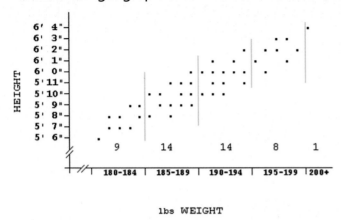

Figure 115: Process for Converting a scattergram to a bar graph.

Would it be better to have intervals to show the data? For example along the weight axis: 180-184; 185-189; 190-194; above 194; etc. Generally, the interval data is easier to read as long as it is correctly labelled.

Consider: If you were to include all the frequency intervals even where there is no data collected, the graphic could be so cumbersome that it would be useless to show the data you did collect. Intervals would be better because the intervals are written horizontally. Whereas the individual weight were written vertically which makes the labeled items more difficult to read. Using intervals enables one to write the information horizontally. However, that would also condense the dots representing the data. Possibly using a bar rather than dots which is not always advisable. Dots suggest that this represents an individual; while a bar suggests a group. **You need to determine exactly what you want to show.**

LINE GRAPHS

There are several forms of line graphs. Instead of bars to represent the frequency, dots are placed on a grid and then they are joined by lines.

Suppose you are working with a third grade teacher. She has three students who have not liked reading and they are now getting special tutoring. She wants to know if, as their skills in reading improve, they will read more books. As there are only three children to be considered, you could use a single graph for each child.

"John's" data to the right; the number of books he has read for each month of the year is recorded. You can tell that as the year progresses, he reads more books. There is a problem with this graphic. Because the line representing the data is black, the horizontal line at "4 books read" does not show up well.

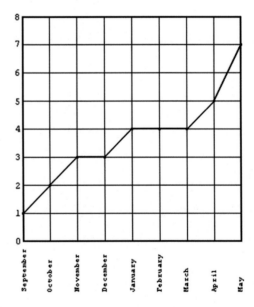

Figure 116a: Line Graph.

Below is a comparison of two line graphs showing the same data and one solution for the problem:

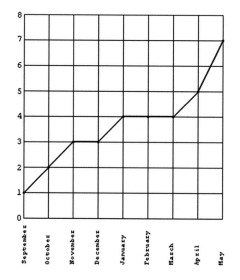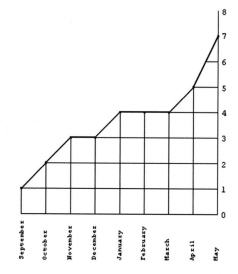

Figure 116b: A silhouette has been used to emphasize data.

The graph on the right has had the grid removed above the line recording the data. By cutting out a portion of the background making the data show as a silhouette, John's data becomes more dramatically obvious. While the graph on the left does "reveal" the monthly total, the one on the right is a style that usually is used to show *cumulative* data (like having a running total).

Someone reading the one on the left could assume that John read one book in September, one in October, one in November none in December (or February and March), one in January, one in April and two in May for a total of 7 books. When using line graphs, check with a friend to be sure it is saying what you think it is!

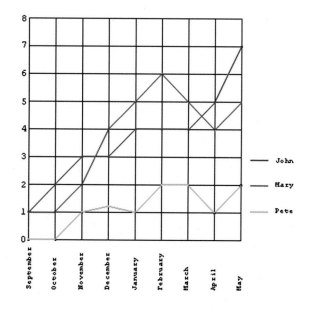

Figure 116c: A graph combining data from the three students.

Let's return to the data from the third grade teacher for three students "John," "Mary," and Pete." It is possible to chart each of the three children on the same chart. But to do this, a different line must be used for each child so that the teacher can tell immediately which line applies to whom. The chart must be labeled to show which line is for which student. John's data has been changed to blue.

The graph shows the month-by-month progress totals for each of the children. The line-colors are kept the same if you have more charts to show other data. (Don't confuse the teacher by changing colors!)

Instead of colors, patterns such as dots or dashes may be used. If you have four sets of data, dashes separated by dots may be used. Just make sure each pattern is clearly different.

The data can be shown as cumulative totals: John read thirty-three books and Mary read thirty-four. The results are close both in numbers and the lines on the cumulative chart, just a slightly different pattern. However, the monthly chart shows the differences in pattern for John and Mary clearly. The cumulative-total chart shows this more clearly than the monthly-total chart does. (This graph would need to be larger in a real report.)

NOTE: If you were to use this chart, an explanation would be needed concerning the ¼ of a book Pete read in December.

It might go like this: *"Pete read only 10 books. The 1 and ¼ books in December may be misleading as Pete read one book in November and another in December. During December he started on a second, but he didn't finish that book until January, so he read the equivalent of one full book in each of the months of November, December, and January."*

An alternative chart would show the cumulative data from the study. An advantage to this type of graph is that it really shows how much John & Mary have progressed and it also shows that Pete has progressed but not as much. [Drawn small to fit in here; the teacher would need a full page-sized version.]

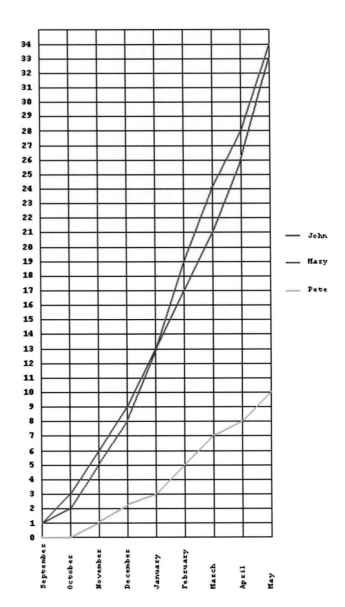

Figure 116d: Line Graph showing cumulative data.

The original question was whether the special reading program for the three who didn't like reading would work; and it did for John and Mary.

To understand Pete's data, you need to know more about Pete. There are some boys who are so in need of action that they cannot sit still long enough to do a lot of reading. Or perhaps the person giving the special reading course expected the children to do reading at home - maybe Pete has a lot of sports activities after school and is too tired to read at the end of the day. Or maybe Pete still doesn't enjoy reading. The extra information helps to explain why Pete hasn't read more.

Some boys are just not turned on by reading: a 10-year old boy was tested (Stanford-Binet reading test) a couple of weeks before the end of the school year and was found to be 18 months *behind* his age level for reading. Three months later three weeks after the beginning of the school year he was tested again. This time he was 18 months *ahead* of his age level. What made the difference? He discovered children's encyclopedias and was excited by the reading he was able to do. Suggest that Pete's teacher provides a large variety of different types of books to see what might turn him on?

Here is another line graph that has a problem. **Both axes** must be correctly labeled or the reader is left guessing.

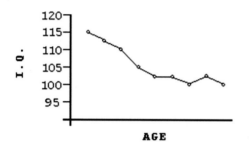

Figure 117: Insufficient data makes a puzzle.

• Although implied, one cannot be sure the age increases from left to right. Or does it? Maybe age **decreases** from left to right?

• What age or ages are being tested?

• Maybe only one person is being tested over a period of time?

• Or does this represent the I.Q.'s of nine different people on the same day? One person's data says something quite different from nine people's data.

BAR GRAPHS:

Bar Graphs are two dimensional, with the vertical and horizontal axes labeled. Using the data in the sorted array or the tally sheet, here are two examples with the one on the left showing the horizontal bars, in the other, vertical bars. ***Note: that the frequency axis is in the same direction as the bars.***

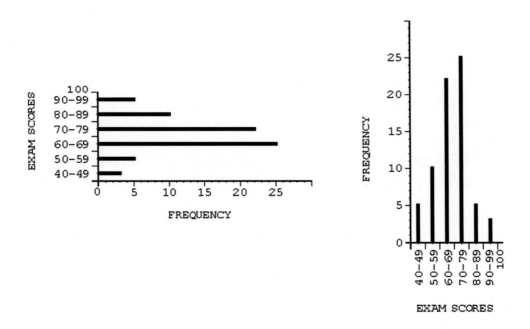

Figures 117a & 117b: Bar graphs with different orientation.

Note that the data is listed as intervals (60-69, 70-79, etc.). Each interval fits exactly between the short bars on the vertical axis (and those on the horizontal axis to the left). Which representation you use depends on the space available and whether you have a preference.

TABLES & ARRAYS:

Tables can be used for any of the three types of data considered in this course. A table is fairly easy to create as most word processing software has a built-in way of creating tables. You just have to know what you want displayed. There have been several tables used in the course materials. Here is one that shows the difference between the three types of data:

DATA TYPE	NATURE OF THE DATA	DESCRIPTION
Nominal	Categories	Classification by attribute
Ordinal	Ranked Order	An ordered sequence
Interval	Numeric Values	Values falling in a continuum

Figure 119: Types of Data.

The most commonly used graphics to illustrate one's research report are those that show differences in frequency (interval data). Having obtained the data, it needs to be displayed in a form that can be easily interpreted. A table is useful for this. Don't forget, doing research is an effort to not only enlighten yourself but also to enlighten others through your report. Illustrations in the text of your report will enable future readers to more clearly understand what you were trying to do, how you did it, and your conclusions from the results.

This Table is acceptable, if there is only a small amount of data. However when you have larger quantities of information, a table may not be adequate.

Hour	gals per min
1	81
2	76
3	51
4	65
5	56
6	73
7	85
8	91

Figure 120: A table showing an array of data.

The data you collect is referred to as *raw* data before you do anything to it. *Tabulation* refers to the arrangement of data in columns or tables or in an array. When all the data is collected it is usually unsorted and the first activity is to sort it into some kind of order.

UNSORTED ARRAY:

Generally, "raw" data is placed in an array without organizing it first. Unsorted arrays contain raw data. The number of rows and columns doesn't matter as long as all data is included. Following data could have been written as a 7x10 or a 10x7 array.

18, 21, 23, 22, 17, 25, 27, 22, 27, 19, 20, 28, 18, 19, 21, 20, 26, 25, 19, 24, 18, 17, 21, 20, 21, 18, 19, 21, 21, 25, 20, 22, 23, 19, 20, 21, 20, 26, 18, 19, 20, 25, 19, 21, 20, 17, 22, 20, 18, 19, 26, 22, 20, 22, 20, 21, 19, 18, 17, 24, 25, 19, 19, 28, 26, 23, 23, 24, 22, 24

If the data is in a, for example, 10x7 array, it looks "tidier" and more easy to read.

18, 21, 23, 22, 17, 25, 27, 22, 27, 19,
20, 28, 18, 19, 21, 20, 26, 25, 19, 24,
18, 17, 21, 20, 21, 18, 19, 21, 21, 25,
20, 22, 23, 19, 20, 21, 20, 26, 18, 19,
20, 25, 19, 21, 20, 17, 22, 20, 18, 19,
26, 22, 20, 22, 20, 21, 19, 18, 17, 24,
25, 19, 19, 28, 26, 23, 23, 24, 22, 24

However, because the array is unsorted the data isn't easily accessible. When sorting data with as many items as this array, it is easier to sort using a tally sheet.

Age	Tally	Total
17	IIII	4
18	ⵏⵀ II	7
19	ⵏⵀ ⵏⵀ I	11
20	ⵏⵀ ⵏⵀ II	12
21	ⵏⵀ ⵏⵀ III	13
22	ⵏⵀ ⵏⵀ	10
23	ⵏⵀ I	8
24		0
25	ⵏⵀ	5
		70

Figure 121: Tally sheet.

TALLY SHEET:

When you have unsorted data, using a *tally sheet* is helpful in counting the frequency, that is, the number of occurrences in a particular class or class interval.

To make a tally, you use a single vertical stroke to count, but when you get to the fifth count, you mark across the previous four with an angled line to indicate there are five. Then when getting the totals, count the groups of five and add any single lines to the total.

The tally sheet sorts the data for you. A tally sheet of the data in the unsorted array looks like this (note that a zero frequency is possible):

SORTED ARRAY:

Generally an array is sorted in ascending order, that is from lowest to highest but it is acceptable to sort in descending order (highest to lowest). As you can see, now the data has been sorted, it is easier to determine the frequencies of each value.

17, 17, 17, 17, 18, 18, 18, 18, 18, 18,
18, 19, 19, 19, 19, 19, 19, 19, 19, 19,
19, 19, 20, 20, 20, 20, 20, 20, 20, 20,
20, 20, 20, 20, 21, 21, 21, 21, 21, 21,
21, 21, 21, 21, 21, 21, 21, 22, 22, 22,
22, 22, 22, 22, 22, 22, 22, 23, 23, 23,
23, 23, 23, 23, 23, 25, 25, 25, 25, 25

In a small study, the amount of data may be fairly simple to deal with, but when we are dealing with data obtained from 100 or more observations, it becomes rather tedious to deal with each data point or observation when calculating various statistics. That is why we use a *frequency distribution* for the data. Suppose you are working with a particular professor with 70 students and the following exam data:

SCORES	TALLY	# OF STUDENTS
40-49	|||	3
50-59	|||||	5
60-69	||||| ||||| ||||| ||||| |||||	25
70-79	||||| ||||| ||||| ||||| ||	22
80-89	||||| |||||	10
90-99	|||||	5
	Total	70

Figure 122: a tally sheet used to record data.

make a real comparison of improvement one would need to know each individual's scores. This graphic, however, is intended for calculating the means for both exams, and to compare the means.

The scores are shown with interval data, and the tally has been summarized in the third column.

Generally, a professor will want to compare results from two exams. In the following table are the scores of the mid-term and the final exam.

Intervals have been used, each with a range of ten. A cursory look at the scores suggests that there have been some improvement, but to

MID-TERM EXAM	# OF STUDENTS	FINAL EXAM	# OF STUDENTS
40-49	3	40-49	1
50-59	5	50-59	7
60-69	25	60-69	19
70-79	22	70-79	28
80-89	10	80-89	8
90-99	5	90-99	7
Total	70	Total	70

Figure 123 - Interval data saves space.

NUMBER LINE ARRAYS

A number line is a horizontal line containing some part of the set of whole numbers (from zero at the left end of the line to infinity along the right). The appropriate markers are set along the line and labeled. Or it may represent the integer number set around the zero point. Negative values are to the left of zero and positive values to the right. The integer number line goes to negative infinity and also to positive infinity.

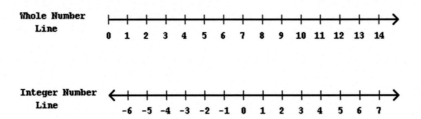

Figure 124a: Whole and Integer Number lines.

Note: the Integer number line may also be referred to as the *real number line*. The difference is that the integer number line contains only the points where the numbers are labeled whereas the real number line contains all the points in between (fractions, decimal values, etc.). Strictly speaking the whole and the integer lines should be a series of labeled dots but it is simpler and acceptable to just draw a line.

Example: An ordered array may be represented graphically on a number line. In this example, "X" has been used to represent the data points, but any suitable symbol may be used: ●, ○, **0**, etc. The raw data array would be plotted as the number line below the raw data:

<u>Raw data</u>   12, 18, 18, 18, 18, 18, 19, 22, 22, 22, 25, 27, 29,
                31, 32, 33, 35, 36, 36, 37, 38, 38, 43, 44, 45

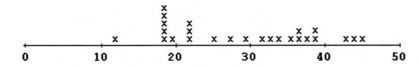

If there was sufficient data, the number line array would begin to look like a bar graph.

Figure 124b: 'x' is used to show the data points.

There is another type of array where data can be placed along a number line. This would probably be best used when there are fairly small frequencies. But it is very useful in displaying *z-scores*. The dashed line at the top of the graph contains the actual data: 1, 2, 3, 5, 6, 8, 8, and 10. The mean is on the center vertical line (5.375). The solid line along the bottom shows the standard deviations of 0 (at the mean) and ±0.5 **S**, ±1.0 **S** and ±1.5 **S**.

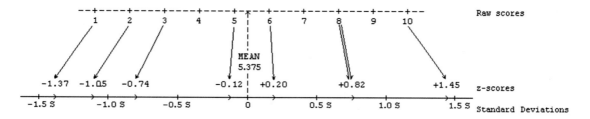

Figure 125: A method for comparing raw scores with z-scores.

METHODS FOR ENTERING DATA IN A TABLE ARRAY:

Below, with the same data: the frequency table on the left shows where the information has been entered row by row {horizontally, that is: (→).} People who speak English tend to read from left to right and then down to the next line. However, personal preference may be to enter the data downwards. The array on the right has been entered column by column {vertically, that is: (↓)}:

ENTERED HORIZONTALLY

12	18	18	18	18
18	19	22	22	22
25	27	29	31	32
33	35	36	36	37
38	38	43	44	45

ENTERED VERTICALLY

12	18	25	33	38
18	19	27	35	38
18	22	29	36	43
18	22	31	36	44
18	22	32	37	45

Figure 126: entry is a matter of personal preference.

HISTOGRAM:

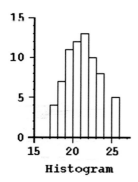

Histogram

A histogram is similar to a bar graph but the data is in columns that are touching each other. In the bar graph, the columns are separated. The height of the column represents the frequency. Note: the horizontal axis represents the nominal values of age (15 through 26). The vertical axis represents the frequency for each age.

Figure 127b: Histogram with a zero frequency.

Sometimes the frequency for an item of data is zero. The value of zero (for the nominal label, 24) has no column Note: In this graphic, the width of the column is the difference between one nominal amount and the next (for example, the fourth column sits between 20 and 21 and represents the "age" 20). This means that the middle of the top of each column provides and can be marked with a midpoint. Midpoints are used to create a frequency polygon and a smooth curve.

THE FREQUENCY POLYGON & THE SMOOTH CURVE:

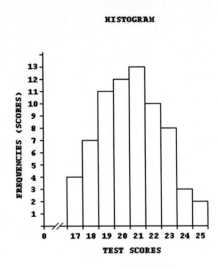

Figure 127a: A histogram.

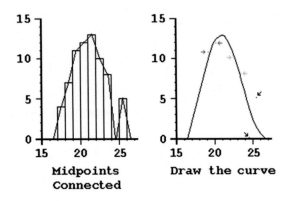

Figure 128a -Histogram & smooth curve.

To draw the frequency polygon or a curve, you connect the midpoints. Either graph will be useful in illustrating your research report. The mid points are indicated by the top dot of each bar. To draw the polygon or curve, the end points (zero and 100) *must* be included. To draw the curve, you smooth the data by drawing the curve between the closest consecutive pairs of midpoints (red, green, and black arrows). It isn't necessary to show the zero frequency at 24 but note the lower black arrow indicates that zero point.

Here is another example of converting a bar graph to a smooth polygon: In this case because there is a sharp peak in the polygon, this curve also has a fairly sharp peak. The curve is skewed (not symmetrical).

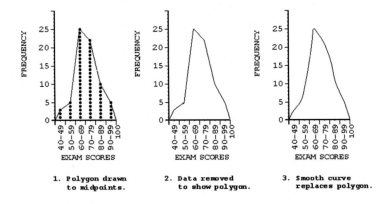

Figure 128b Bar Graph to polygon to curve.

SCATTER DIAGRAMS (aka SCATTERGRAMS):

The scattergram is very useful when two variables are being compared to determine if there is a relationship between them. For example, there is a relationship between height and weight. The relationship is shown by the type of "spread" of the data points. If the data is spread as in the height/weight diagram, then there is a direct relationship (linear). In words: there is a "direct" relationship between height and weight (the taller a person is the more likely they are to weigh more. The problem with this graphic is the labels on the axes: what does "tall" and short mean? And what does "light" and "heavy" mean?

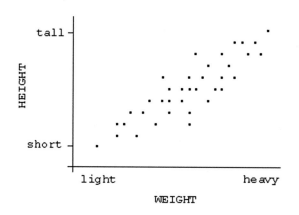

Figure 129a: Inadequate labelling of the two axes.

Can data be deliberately skewed enough to be intentionally misleading? YES! On a field trip to Africa, the geology team took several photos. They placed the tallest person (6'4") beside two of the shortest pigmies (less than 3'2") and later the shortest person (5'4") beside the tallest pigmies (3'10" and 3'9") before taking their photos! Because one tends to assume that the pigmies were all the same size, the contrast between the tallest and shortest persons was greatly exaggerated much to the joy of the photographer!

When either or both axes are not properly labeled, the best than can be said of Figure 129a is that is it misleading. Looking at the correctly labeled graphic below:

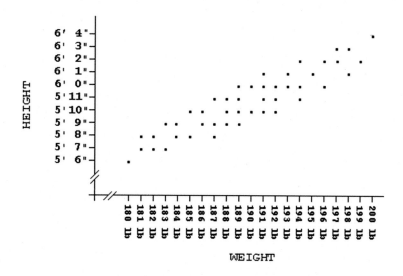

Figure 129b: Better labelling. BUT...

The author of the report must make a decision. Should the weight axis be labeled as here with individual weights or as interval data? Since the data is graphed as a scattergram, it would be better to label as shown. If the weights were to be separated into intervals, a line drawing would be more appropriate. There is another problem with that as the whole illustration would need to be changed. Then interval weights in a line drawing really should be compared with the frequency and the use of a scattergram comes into question. It would be possible to use intervals with both height and weight, but the illustration would be less self-explanatory and because it is more complex it is also difficult to draw. KISS PRINCIPLE: KEEP IT SIMPLE STUPID!

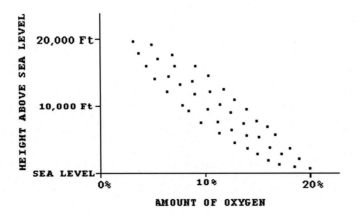

Figure 129c -Scattergram comparing oxygen above sea-level.

A scattergram, with fewer labels can be used if preferred; if, for example, the data spread is showing a comparison of the height above sea level and the amount of oxygen. Note, this relationship is an "inverse" relationship. In words: the greater the height above sea level, the lower the amount of oxygen.

Note there is more detail in labeling the two axes in the diagram. This is preferable to the height/weight chart in Figure 129a.

If the data points are in a straight line (not shown), then the relationship is perfect. If the data points slants upward toward the right (/), it is a direct or positive relationship. If the data points slants up toward the left (\), then it is an inverse or negative relationship.

There are three other named relationships. If the data points are scattered in an oval shape, but are more tightly arrayed than the above scattergram, then there is a moderate relationship: *direct or positive* if the data points rise toward the right of the diagram, *inverse or negative* if the data points rise to the left. When the data points are scattered in a ball, there is said to be no relationship.

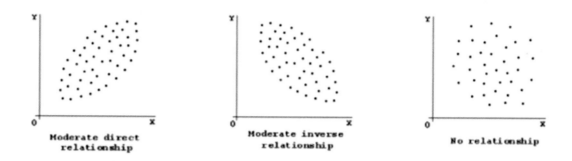

Figure 129d - More typical distributions.

FLOW CHARTS:

When you are trying to show the choices in selecting, for example, the appropriate test for a particular type of data, a flow chart is probably most useful such as the one at the right for certain Chi-Square tests. {NOTE: At this time you do not need to worry about the information on the flow chart; this chart is just here to illustrate the appearance of a flow chart. The information on it was explained in the unit dealing with nominal data.}

Flow charts may "flow" in any direction. Because people in English-speaking countries tend to read to the right and downward, this chart flows down

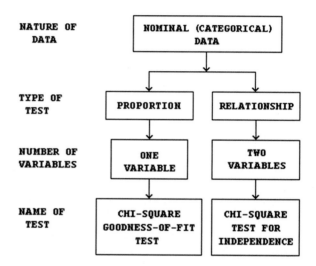

Figure 130a: Using arrows to indicate direction of flow.

This chart, on the other hand flows UP. While there is no definite rule for the direction of flow, most people would be uncomfortable with charts flowing in this direction.

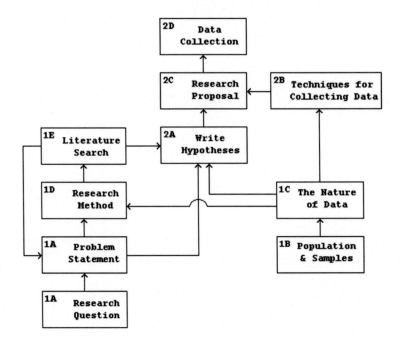

Figure 130b: Chart that flows UP.

Figure 120b has been inverted so it appears to be "right-side-up" However, having cross-over lines may also be confusing. The semi-circular arch shows how to make a cross-over line between 1C and 1D. On the next page is a simplified version of this chart.

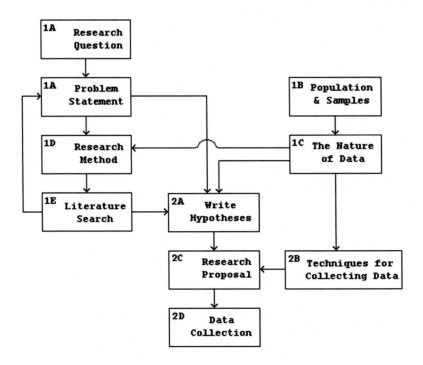

Figure 130c: "Right" side up but cross-over arch may be a problem.

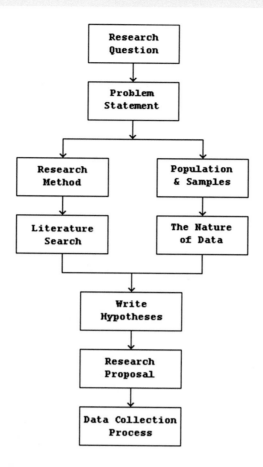

Figure 130d: Simplified and improved flow chart.

Simplifying the chart makes it much easier to read.

RECCOMMENDATION: You may need to rework any flow chart one or more times in order to create the best representation.

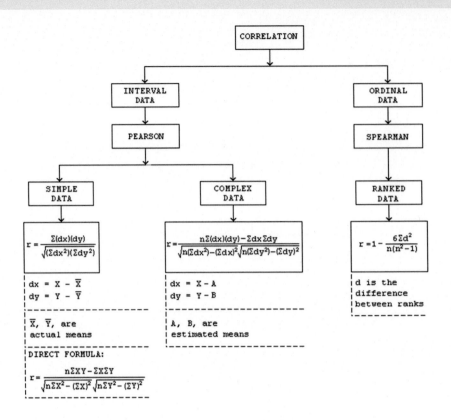

Figure 130e: a chart with branches & explanatory information.

Additional information may be added to the chart, if it is needed for better understanding the information. To show that this explanatory material was NOT part of the flow chart itself, dashed lines were used.

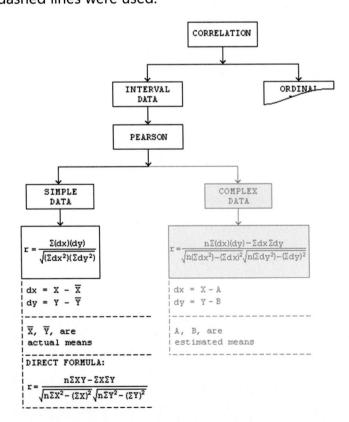

Your chart may have the complete picture on it:

But there may be times when you wish to use the same chart but need to emphasize one of the sections.

Here are two methods for doing this to de-emphasize the unneeded information. One is to delete part of the box ("Ordinal") leaving just enough to show what the box originally referred to.

The other is to "grey out" one (or more) portions of the part (s) of the chart that need to be de-emphasized.

TREE DIAGRAM:

All statistical tests are based on probability. One type of probability is called *conditional probability*. This means that it may be necessary for a certain event to occur before a second event is likely to occur. For example, a researcher wishes to estimate the sex (male or female) of the first three possible offspring for one of her experimental animal.

This is the table she came up with, called a *sample space* showing the possibilities for the first three offspring:

M M M	All offspring are male.
M M F	Two males followed by a female.
M F M	One male followed by a female then another male.
M F F	One male followed by two females.
F M M	One female followed by two males.
F M F	One female followed by a male then by a female.
F F M	Two females followed by a male.
F F F	All three offspring are female.

Figure 131a: a sample space shown in a table.

The condition for the first four items in Figure 131a is that the fist offspring must be male. The condition for the last four is that the first offspring must be female. The conditional probability for the last two items: If there are two females for the first two offspring then what is the probability of the third being male. In this case the condition is "two females first." There are two items that fill this condition, but only one outcome has two females followed by one male (F F M). Each of these outcomes is referred to as an *event*. A simple event is a single outcome of a sample space. Each *possible outcome* in the offspring sample space is a *simple event*.

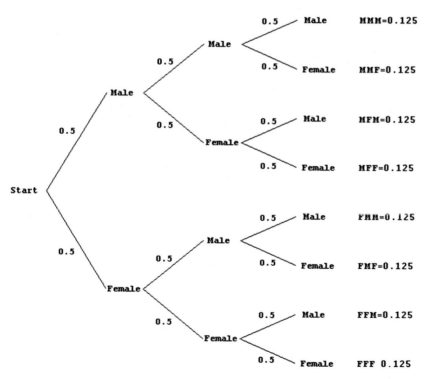

The simple events form the basis of the Tree Diagram. The outcomes at the end of each branch are found by reading along a particular branch.

In the diagram, the first event will be either male or female. With the probability in each case of 0.5. If this event is male then the next offspring

Figure 131b: Tree Diagram of offspring probabilities.

could be male or female. Again the probability, because there are two possible events, is 0.5. The probability of male followed by male is 0.5 x 0.5 or 0.25.

The probability of male, male and male is 0.5 X 0.5 X 0.5 or 0.125.

The Tree shows all possible outcomes with the probability of the simple events listed at the right end of the tree's branches.

The probability of a simple event (the end point of each branch) occurring is the quantity 1 divided by the total number of simple events possible. In this example, the probability of any **one** of the simple events occurring is 1 in 8, or 1/8, or 0.125 or 12.5%.

If the researcher wishes to know how likely it is that the animal will have at least two male offspring. You can read from the chart which outcomes have at least two males. These outcomes are:

M M M	**M M F**	**M F M**	**F M M**

To determine the probability of one of these four events happening, you divide the number of events by the total possible events, that is, 4 out of 8 (4 ÷ 8) or ½ or 50% or 0.5. So the probability of having at least two males is 50%.

TO CREATE A TREE DIAGRAM FOR A RESEARCH REPORT:

The intent is to get each branch equidistant so the tree is neat in appearance. Suppose an elementary teacher wanted to map the outcomes of weather **for one week (5 days)** in a place where in the summer, the weather is only **sunny** or **cloudy.**

1. Decide how you will order the possibilities (sunny, cloudy OR cloudy, sunny); whichever order you choose **must** be maintained throughout.

2. Make a rough sketch of all the outcomes from start to each branch end.

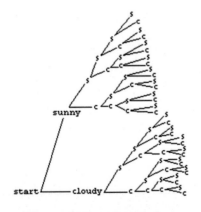

Note: the number of outcomes double as you go out on another set of branches. There are five sets of branches (for the week) represented by dividing lines (<). So there are 32 outcomes total on the 5th set of branches.

It isn't necessary to make sure the various branches are correctly positioned in your rough sketch. The purpose of the sketch is to determine how many branches are needed in the final graphic.

Figure 131c: Sketch for a Tree diagram of weather.

3. Begin the finished draft, starting at the **end** of the branches and work backwards to the "start" point. You may want to use graph paper.

Five steps are needed, and you need to work from the right at the end of the branches.

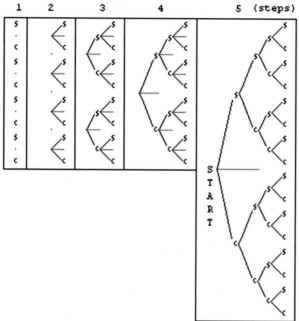

Figure 131d: Steps to creating a tree diagram.

1. Make a vertical list of the items at the end of the branch. In this example, there are thirty-two outcomes (but only part of the tree is drawn here).

2. Be sure they are equidistant vertically and mark the half-way point between each pair of outcomes on the list.

3. Draw slanting lines (black) from the left end of a line drawn from the half-way point (red) to the outcomes (S & C) for each pair of items.

4. Repeat item 3.

5. Repeat until you can show the "Start" point of the tree.

In this illustration, the horizontal dimension is tight. When you draw your own, increase the length of each slanting line and use a smaller angle.

APPENDIX C: CHANCE, PROBABILITY & RANDOMNESS

DEFINITIONS THAT APPLY TO DATA:

CHANCE:

The term "chance" as used in statistics may have a frequency interpretation or philosophical interpretation. It has to do with things unknown and ways of *estimating* likely the occurrence of some event, e.g., the number of people lined up in the bank at a cashier's station at any given moment in time. First, the researcher would need to make observations to determine the pattern of occurrences, identify the set of all possible outcomes and the subset of outcomes of particular interest.

PROBABILITY:

Probability measures how likely something will occur, the chance of something specific occurring. It is also possible to measure if the occurrence of one event may greatly affect the occurrence of another.

Probability has to do with things unknown and a way of estimating the likely occurrence, for example, when and how many cashiers in a bank should be on duty? What is the likely number of people waiting for a cashier's assistance at any given time of the day? Probability enables the scheduler to estimate the number of cashiers needed at any given time of day.

The scheduler needs to take observations to determine the pattern and identify the set of all possible outcomes, and the subset of items of interest, then estimate the likelihood of how many customers are likely to need a cashier's assistance. Another question that will need to be answered is estimating whether there are differences between the days of the weeks.

There are five approaches to the concept of probability.

CLASSICAL APPROACH.

When a balanced die is rolled, there is an even chance that any one of six sides may turn up. When using the classical approach, it is customary to assign probabilities of 1/6 to each side on the assumption that each side has an equal chance of turning up no matter how often the die are rolled.

RELATIVE FREQUENCY.

If the die is not perfectly balanced, it is illogical to suppose that each side has an equal chance of turning up. By rolling the die a large number of times one can record the frequency with which each side turns up. The relative frequency becomes the measure of the probability of any one side turning up. This may be called an empirical approach, that is, based or verifiable by observation or experience rather than theory or pure logic.

SUBJECTIVE PROBABILITY

In some circumstances it is impossible to assign a specific probability (fifty people are applying for three jobs, but one cannot assume that "my chance of getting the job is 1 in 17" since there are possibilities the only a portion of the applicants are qualified, "I" may not be, and a number of other factors may be involved). Nor can relative frequency be calculated (one cannot get and quit a job a large number of times to determine the probability of success in obtaining a job). One can however pull together the information available and make a **subjective judgment** of the probability of success.

CUMULATIVE PROBABILITY.

This type of probability is probably best illustrated by considering a coin toss experiment. Suppose you plan to toss the coin twice, and wished to determine the probability of one or less heads being tossed. One or less means that you either get one head or you get no heads (tails). The answer to this is a cumulative probability. It is the probability that the toss will produce no heads **plus** the probability that the toss will produce one head. The possible outcomes of tossing the coin twice are:

First toss: 0 heads or 1 head (0.25 for 0 heads, 0.25 for 1 head).

Second toss: 0 heads or 1 head.

On the first throw, when you combine the results, you have one chance of tossing 0 heads.

On the second throw: If you have already tossed a head, then you have a chance of tossing no heads the second time which gives you one head total. Again, if you have tossed a head and then toss a second head, that gives you two heads total. That means there are two chances of throwing one head and one chance of throwing two heads. The table below summarizes the cumulative probabilities and how these are related to probability:

Number of heads	Probability	Cumulative Probability
0	0.25	0.25
1	0.50	0.75
2	0.25	1.00

Figure 132: Probability & Cumulative Probability.

CONDITIONAL PROBABILITY:

This is the probability that, given that event A has occurred, event B will also occur. That is, the occurrence of event B is conditional on the occurrence of event A. The conditional probability of B given A is equal to the probability of the intersection of A and B divided by the probability of B.

If A & B are mutually exclusive, the probability of their intersection is zero. Then the conditional probability of B given A is also zero. *It is necessary for there to be some overlap or interaction between A and B for the conditional probability of B to be other than zero.*

ELEMENT:

The data points within a set **A** or a set **B**. **A** may contain the elements 1, 2, 3, 4, so **A** = {1, 2, 3, 4} and **B** may contain the elements 3, 4, 5, and 6, so **B** = {3, 4, 5, 6}.

INTERSECTION (∩):

The intersection between **A** and **B** is **C**, where **C** contains the elements that are common to both **A** and **B**. With the above two sets the intersection is 3, and 4, that is, **C** = {3, 4}

UNION (U):

Given two sets of data **A** and **B**, **A U B**, that is, "**A** union **B**" means that the two sets **A** and **B** have been combined (similar to the process of addition).

EVENT:

Events are possible outcomes of an experiment, the outcomes being theoretical, using probability, not based on observations. For example, in a laboratory, the researcher is looking for information on the possible sex of the first three offspring of a particular animal. The offspring can be either M (male) or (F) female. One *possible* outcome is that all three of the offspring will be M (male). This is conjecture based on probability of occurrence not a result of observation. You might say that the researcher is anticipating the possible outcomes of the births before they take place. He is interested in the sequence of the birth of the offspring.

SAMPLE SPACE:

"Sample Space" refers to all possible outcomes of an experiment. For example: the following set of outcomes is a sample space with a total of eight (8) possible outcomes for the first three offspring:

EVENT X	P(X)	OUTCOME DESCRIPTION
MMM	0.125	All three are male.
MMF	0.125	Two males followed by one female.
MFM	0.125	One male, then one female, then one male.
MFF	0.125	One male followed by two females.
FMM	0.125	One female followed by two males.
FMF	0.125	One female, then one male, then one female.
FFM	0.125	Two females followed by one male.
FFF	0.125	All three are female.

Figure 133: Sample space with probability.

SIMPLE EVENT:

"Simple Event" is a single member of a sample space or a single outcome of an experiment.

- Each line (possible outcome) in the above sample space in Figure 133 is a simple event.

- The probability of a simple event occurring is the quantity 1 divided by the total number of simple events possible. In the above example, the probability of any one of the simple events occurring is 1 in 8, or 1/8, or 0.125 or 12.5%.

COMPOUND EVENT:

A *Compound Event* is a group of one or more simple events in the sample space. The probability of a compound event occurring is the sum of all possible simple events that fit the defined compound event divided by all possible events.

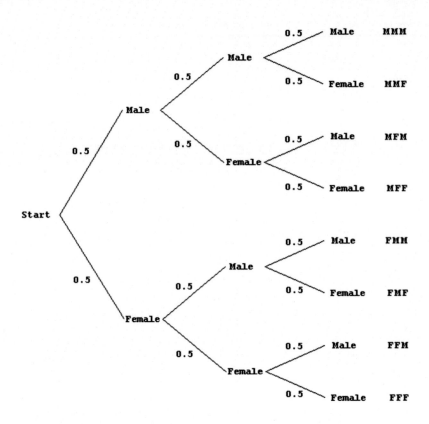

Figure 134: Tree with Offspring illustrating a compound event.

Suppose we are looking at the possibility of two of the three offspring being male.

- Example: List all possible outcomes in the sample space:

MMM, MMF, MFM, MFF, FMM, FMF, FFM, FFF

- List the outcomes of the event "There are at least two males." Note: the use of the term "at least" allows us to include the outcome of MMM. If the word had been "only" then MMM would be eliminated.

MMM, MMF, MFM, FMM

- The probability of this compound event occurring is 4 out of 8 or ½ or 0.5 or 50%.

What about the probability of at least one male and at least one female? The 'and' in that statement excludes the possibility of three males or three females. 'At least' means there could be two males or there could be two females. But to include MMM and FFF, the statement would have to include "at least one male OR at least one female."

- Looking at the tree diagram, there are six possible outcomes with at least one male and one female.

MMF, MFM, MFF, FMM, FMF, FFM

The probability of this compound event occurring is 6 out of 8 or 3/4 or 75%.

INTERSECTION OF EVENTS (∩):

An intersection is the combining of two or more events, in such a way that only common events are included. A Venn diagram is best for illustrating the intersection of the outcomes of at least two males:

Sample space: MMM, MMF, MFM, MFF, FMM, FMF, FFM, FFF

EVENT A: MMM, MMF, MFM, FMM
EVENT B: MMF, MFM, MFF, FMM, FMF, FFM

Sometime it helps to "spread and match" the common elements:

EVENT A: MMM, MMF, MFM, FMM
EVENT B: MMF, MFM, MFF, FMM, FMF, FFM

THE INTERSECTION: MMF, MFM, FMM

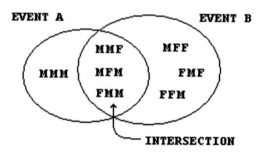

Both the boundary of event A and the boundary of event B include and enclose the intersection items.

Figure 135: Venn diagram showing an intersection.

COMPLEMENT OF EVENTS:

Complement is a word derived from "complete". **A'** is referred to as *A Prime* and contains all the simple events in the sample space **not contained** in **A**, in other words, the *complement* of **A** is **A'**. By definition **A + A'** = the full set of events in the sample space.

[Note: If U = the union of two sets, A U A' is the set of all possible events in that sample space.]

COMPLEMENT

EVENT A			OF A (A')			A U A'		
M	M	M		-		M	M	M
M	M	F		-		M	M	F
M	F	M		-		M	F	M
	-	M	F	F	-	F	F	M
F	M	M		-		F	M	M
	-		F	M	F	F	M	F
	-	F	F	M		F	M	F
	-	F	F	F		F	F	F

MUTUALLY EXCLUSIVE EVENTS:

Two events that have no simple events in common, that is, their intersection is the empty set (∅). [Note: **A** & **A'** are mutually exclusive.]

Example: where all offspring are the same sex.

EVENT A			EVENT B		
M	M	M			
	-			-	
	-			-	
	-			-	
	-			-	
	-			-	
	-			-	
	-		F	F	F

UNION OF EVENTS:

Example 1 (mutually exclusive):

EVENT A			EVENT B
M	M	M	
-			--
-			--
-			--
-			--
-			--
-			--
-			F F F

Union (**A ∪ B = {**M M M} + {F F F} = {M M M F F F}

The probability of the union of mutually exclusive events occurring is the sum of the probability of each event (the addition rule). The probability of the union occurring in the example is 1/8 + 1/8 or 1/4 or 25%.

Example 2 (not mutually exclusive):

EVENT A			EVENT B			INTERSECTION (A + B)		
M	M	M	-			M	M	M
M	M	F	M	M	F	M	M	F
M	F	M	M	F	M	M	F	M
	-	M	F	F		M	F	M
F	M	M	F	M	M	F	M	M
	-		F	M	F	F	M	F
	-	F	F	M		F	F	M

The probability of union occurring where events are *not* mutually exclusive is the sum of their separate probabilities *minus* the probability of their intersection occurring. The probability of the union occurring in Example 2 is 4/8 + 6/8 - 3/8 or 7/8 or 87.5%

OTHER EXAMPLES OF PROBABILITY USING A PACK OF 52 CARDS:

- The probability of selecting a queen from a pack of 52 cards is 4/52 (4 queens in the pack).

- The probability of selecting a heart is 13/52 (each suit has 13 cards).

- The probability of selecting a queen of hearts is 1/52, the intersection of queen $\cup$ heart (only one queen of hearts).

- The probability of selecting a queen *or* a heart. This is the union of queens and hearts: 4/52 + 13/52 - 1/52 = 16/52 or 4/13 (must subtract 1/52 otherwise the queen of hearts is used twice).

- The probability of selecting neither a queen or heart, that is the complement of queen or heart (all **other** cards in the deck):   1 - 4/13 = 9/13 (the '1' representing the full set of cards or 52/52 = 1)

OTHER DEFINITIONS THAT APPLY TO ANALYSIS OF DATA:

BIASED/UNBIASED SAMPLE STATISTICS:

- Correction factor from a biased estimate of population variance to an unbiased estimate (i.e. divide by n-1 rather than n)

INFERENTIAL STATISTICS:

- Trying to predict or estimate the population parameters (mean, variance) from the statistics of the sample you are using.

- When dealing with sample statistics must always control for chance—concern: that one didn't get a true cross-section.

- If dealing with population statistics, do not have to test for significance because no chance is involved (but must be sure that one has the whole population represented in one's data)

RANDOM SAMPLING:

- Challenge -- determine the size of the sample in order to get a good representation of the population without getting the whole population, that is, to control chance.

- Determine size of population (infinite or non-infinite)

- Determine the amount of confidence you wish in the results; (how important is it that you get a true result?)

- Define the percent of occurrence of the attribute in question in the population.

- Define the precision of your result (how accurate you want the answer to be, and how confident you want to be in the result).

APPENDIX D: COMBINATIONS AND PERMUTATIONS:

WHAT'S THE DIFFERENCE?

In English we use the word "combination" loosely, without thinking whether the order of things is important.

"My fruit salad is a combination of apples, grapes and bananas."

> We don't care what order the fruits are in, they could also be "bananas, grapes and apples" or "grapes, apples and bananas," it's the same fruit salad.

"The combination to the safe was 472."

> Now we do care about the order.

> "7-2-4" would not work, nor would "2-4-7". It has to be exactly 4-7-2.

So, in Mathematics we use more precise language:

> If the order doesn't matter, it is a **Combination**.

> If the order does matter it is a **Permutation**.

> > → A Permutation is an ordered Combination.

We call this a "combination lock." But the order does matter!

So, we should really call it a "Permutation Lock!"

For more, see:

http://www.mathsisfun.com/combinatorics/combinations-permutations.html

Figure 136: A combination lock or a "permutation lock"?

This page left blank to maintain pagination.

APPENDIX E. THE PROCESS FOR WRITING DISSERTATIONS OR A THESIS

THE DISSERTATION WRITING PROCESS.

A dissertation is a series of chapters that are designed to *present your research work* to professional academics. A Thesis is a similar, generally simpler version of a dissertation.

You will be required to defend it. The committee you will be defending it before should consist of your advisor and several faculty members who are experts in your field or closely related to that field (i.e., those who have sufficient knowledge of the subject of your dissertation to be familiar enough with the subject to critique your dissertation). NOTE: a Thesis is basically similar to a Dissertation, just likely to be briefer. The report you write for this class may consist of paragraphs instead of chapters, and you will defend it before your class members.

CHAPTER - 1

The first chapter is an introduction to the area of research that you have chosen. Here is where you explain the issues you are concerned with. Your introduction must accomplish two critical tasks. First, you must explain the problem that you have identified, and second, you must clearly verbalize the research question that focuses your research. The introduction is also expected to prepare the reader for reading and so it summarizes your entire project: your research question, the methodology you chose for framing your research, a brief summary of the research results, and the conclusions you have arrived at.

CHAPTER - 2

The next chapter is expected to demonstrate your familiarity with the issues. Here you must make reference to the research that is already published, and critically analyze the literature. In this chapter you should give details about the research problem and refine your research inquiry so that your purpose is specific and focused.

CHAPTER - 3

The methodology chapter is where you verify for your audience that your research method is valid and supported – using appropriate quote(s) may be helpful. Consider these questions: Why did you choose this method, and how is it appropriate for what you want to explore? What will this method provide you? It may be a quantitative method, providing you with statistical information, or it may be a qualitative method that is analytical, or ethnographic. You might have interviews, or you might be analyzing poetry. Maybe you've chosen to combine several methods. Your methods are specific to your research interests and your objectives and need to be understandable.

CHAPTER - 4

After this, you need to combine everything you learned by discussing your research results. Here is where you examine the data closely, and explain what your analysis has revealed. Whatever your research question is designed to explore, this chapter is where you respond to that question; your reasoning here will help support the validity of your research.

CHAPTER - 5

To complete your work, you have to write a conclusion. The conclusion is similar to your introduction: it is a summary of your work. It is where you re-state the problem, remind the reader of your inquiry, the method, and what your research has revealed. More importantly, the final chapter needs to point towards the future. What is the greater benefit of your research? How does it contribute to your field of study, and what kinds of questions have been unearthed by your research?

FINALLY

In preparing for your defense, after you have finished writing your dissertation, you need to review it as if it were someone else's product. See if you have left something undone or unexplained. Try to anticipate the questions your committee may have for you, and search for answers to those possible question. During the defense, if someone throws you a question you hadn't thought of, ***don't*** say "I haven't thought of that!" It would be better to say, "Let me think about that for a moment!" And try to come up with a reason why you should or should not have thought about. For example: "That's irrelevant to my study because..." or "That's a question that would be of value to consider if further research is done in this area..."

Be alert! You are ***defending*** your dissertation. Be prepared for the committee to tell you to rewrite some of your work. But if you make a good defense, hopefully they will pass your work and congratulate you on getting your doctorate or that your thesis has passed.

Printed in the United States
By Bookmasters